The international politics
of biotechnology

MANCHESTER
UNIVERSITY PRESS

The international politics of biotechnology

Investigating global futures

edited by Alan Russell and John Vogler

Manchester University Press

Manchester and New York

distributed exclusively in the USA by St. Martin's Press

Copyright © Manchester University Press 2000

While copyright in the volume as a whole is vested in Manchester University Press, copyright in individual chapters belongs to their respective authors, and no chapter may be reproduced wholly or in part without the express permission in writing of both author and publisher.

Published by Manchester University Press
Oxford Road, Manchester M13 9NR, UK
and Room 400, 175 Fifth Avenue, New York, NY 10010, USA
http://www.manchesteruniversitypress.co.uk

Distributed exclusively in the USA by
St. Martin's Press, Inc., 175 Fifth Avenue, New York,
NY 10010, USA

Distributed exclusively in Canada by
UBC Press, University of British Columbia, 2029 West Mall,
Vancouver, BC, Canada V6T 1Z2

British Library Cataloguing-in-Publication Data
A catalogue record for this book is available from the British Library

Library of Congress Cataloging-in-Publication Data applied for

ISBN 0 7190 5868 6 *hardback*

First published 2000

07 06 05 04 03 02 01 00   10 9 8 7 6 5 4 3 2 1

Typeset in Photina
by Northern Phototypesetting Co. Ltd, Bolton
Printed in Great Britain
by Bookcraft (Bath) Ltd, Midsomer Norton

Contents

Contents

Figures

Tables

Contributors

Charlotte Bretherton is Senior Lecturer in International Relations and European Studies at Liverpool John Moores University. She has published works on women's policy networking, gender and the external relations of the European Union. She is currently working on gender issues in EU enlargement.

Malcolm Dando is Professor in the Department of Peace Studies at the University of Bradford. He runs the Bradford Project on Strengthening the Biological and Toxic Weapons Convention. Recent publications include, *Biotechnology, Weapons and Humanity* (Harwood Academic Publishers, 1999) and *Biological Warfare in the 21st Century* (Brassey's, 1994).

Hugh Dyer is Senior Lecturer and Director of Teaching and Learning in the Institute for Politics and International Studies at the University of Leeds. His research interests and publications are in the area of international theory, and in the theoretical implications of environmental change.

Robert Falkner is Lecturer in International Relations at the London Centre of International Relations, University of Kent, and Associate Fellow of the Energy and Environmental Programme, Royal Institute of International Affairs. He is currently writing a book on the trade and development policy implications of international biotechnology regulation.

Jez Littlewood is a Ph.D. candidate at the Department of Peace Studies, University of Bradford, undertaking research on biological weapons and arms control. He also serves as a Professional Assistant to the Ad Hoc Group of States Parties to the Biological and Toxic Weapons Convention in Geneva.

Frank Manning is Senior Lecturer in Eukaryotic Molecular Biology in the School of Biomolecular Sciences of Liverpool John Moores Universty. He has over fourteen years research experience as a working molecular biologist. He currently actively researches in the areas of molecular carcinogenesis and toxicology, and the roles of apoptosis in disease. He has authored over thirty five scientific papers, book chapters and abstracts.

Désirée McGraw is a doctoral candidate at the London School of Economics and Political Science. She has worked for the Earth Negotiations Bulletin and currently for the Secretariat of the UN Convention on Biodiversity.

Ian Pownall is Lecturer in the School of Science and Management, University College Scarborough. His research and publications have been on the application of European Union small firm policy with special reference to Irish firms. His work also includes technological issues in international political economy and business.

Alan Russell is Senior Lecturer in International Relations at Staffordshire University. His research interests include technology and International Political Economy with special reference to biotechnology. He is author of *The Biotechnology Revolution* (Wheatsheaf, 1988) and various articles on the political economy of biotechnology.

Joanna Spear is Senior Lecturer in War Studies at King's College, London. Her current research includes work on the proliferation of weapons of mass destruction in the Middle East and the changing political economy of the international defence trade. She has previously published in the areas of arms control and confidence building measures.

Karen Stevenson is Senior Lecturer in the School of Humanities and Social Sciences, Staffordshire University. She has researched sexuality in an age of AIDS, self-identity in modern societies, sexual symbolism and the body and feminist and post-modern theory.

John Vogler is Professor of International Relations at Liverpool John Moores University and chair of the British International Studies Association environment research group. His latest book is *The Global Commons: Environmental and Technological Governance* (John Wiley, 2000).

Simon Whitby is a researcher in the Department of Peace Studies at the University of Bradford, working on the Bradford Project on Strengthening the Biological and Toxic Weapons Convention. He has published widely on anti-crop biological warfare.

Owain Williams completed a Ph.D. in the Department of International Politics at the University of Wales, Aberystwyth. This examined biotechnology under the TRIPs Agreement. Since completion he has worked for a range of NGOs and the WTO in Geneva on the new intellectual property regime for biological resources, biotechnology and patent issues. He has been a freelance consultant preparing trade policy oriented briefings for developing countries.

Abbreviations

AIDS	Acquired Immuno-deficiency Syndrome
ANT	Actor-Network Theory
APEC	Asia Pacific Economic Co-operation
ASEAN	Association of South East Asian Nations
BMD	balllistic missile defence
BTWC/BWC	Biological and Toxin Weapons Convention
BW	biological weapons
CBD	UN Convention on Biological Diversity
CBW	chemcial and biological weapons
COP	Conference of the Parties (of CBD)
CWC	Chemical Weapons Convention
GATT	General Agreement on Tariffs and Trade
GM	genetically modified
GMO	genetically modified organism
HIV	human immuno-deficiency virus
IPR	intellectual property right
IUCN	International Union for the Conservation of Nature
LMO	living modified organisms
MNC	multinational corporation/company
NATO	North Atlantic Treaty Organisation
NBF	new biotechnology firms
NGO	non-governmental organisation
NSF	National Science Foundation
OECD	Organisation for Economic Co-operation and Development
TNC	transnational corporation/company
TRIPs	Trade Related Aspects of Intellectual Property
UN	United Nations
UNDP	UN Development Programme
UNEP	UN Environment Programme

UNSCOM   UN Special Commission
WIPO   World Intellectual Property Organisation
WMD   weapons of mass destruction
WTO   World Trade Organisation

1

Introduction

John Vogler and Alan Russell

Few technologies have excited such widespread interest, or been as controversial, as biotechnology. This is especially apparent when we consider that biotechnology attracted much of this attention from the moment of its incarnation as *new* biotechnology some twenty-five years ago (*old* biotechnology involved the more gradual modification of nature through, for example, plant breeding). The unravelling of the mysteries of DNA provided the essential foundation. It allowed a more precise and speedy means of manipulating plant genetics, but it also opened up the possibility of repeating the exercise with human beings, of cloning organisms and of transferring genes between species so as to manufacture genetically modified organisms (GMOs). The practical and commercial implications were soon evident. New medical diagnosis techniques and pharmaceutical treatments have been developed. Foodstuffs and plants have been genetically modified; fine chemical and bulk chemical production have been influenced; human and animal reproductive techniques have advanced; cloning techniques have emerged and environmental management will increasingly benefit. A new bioelectronics industry is, perhaps, on the horizon. Biological weapons development has also been affected with significant portents for the future. Novel 'designer weapons' are possible alongside safer handling and manipulation of known pathogenic bacteria, viruses and toxins.

A globalised biotechnology industry with a market said to be worth some £70 billion has rapidly emerged. Alongside agro-chemical giants, such as Monsanto and Ciba-Geigy, there were, in 1996, some 584 biotechnology companies in Europe and no less than 1,308 in the United States (Ernst and Young, cited in Consumers Association, 1997: 19). Companies and governments interact in ever-more complicated ways, under conditions of increasing globalisation (Jones, 1995; Stopford and Strange, 1991). Markets are no longer defined as 'national' (Reich, 1991). The boundaries between local, regional and global are becoming blurred as, for example, small firms specialising in niche products look to global marketing strategies. The image of the transnational firm is no longer confined (if it ever was) to the large multinationals. A global market place provides the *raison d'etre* for many small firms that could not survive with a purely local, or even national, reach. While governments and students of international relations (IR) wrestle with the implications of the erosion of national borders and the penetrability of the state, we should not lose sight of the parallel erosion of the 'borders' of large companies or the tenuous relationship they may have with any nation (Badaracco, 1991; Reich, 1991). This latter phenomenon is evident in the increasingly complex pattern of biotechnology-oriented inter-firm links. These are vertical and horizontal, national and transnational, and extend to fundamental research and development in an information age.

Even before many of the applications became a reality, issues of ethics and public safety emerged, often obscured by sensational press reporting of 'designer babies' and 'Frankenstein foods'. In the final years of the twentieth century growing, and on occasion hysterical, public awareness of the reality of genetically modified organisms (GMOs) in food and their release into the natural environment has forced the question of biotechnology into the political mainstream in Britain and elsewhere. Public reaction has ranged from acceptance and apathy in the United States, where genetically modified (GM) crops had become relatively commonplace, to a fearful rejection amongst significant elements of European publics. Intertwined were a complex of sectional and protectionist interests happy to exploit such concerns. In Britain BSE (mad cow disease) and other food-related public health issues had left a fertile soil in which public scepticism towards the reassurances of official science grew. Field trials of GM crops have been uprooted by protesters and supermarkets, and corporations have vied with each other in declaring their products 'GM free'. Public opinion surveys across Europe indicated varying levels of disquiet and distrust, but all attested to the high level of factual ignorance of biotechnology (Consumers Association, 1997: 9–17). As agricultural markets and the transfer of new seed technologies assumed a global scale and as it became clear that the regulation of biotechnology could not be accomplished at a purely national or regional level, indications of the international implications of the technology began to appear. A prominent example was provided by the international regulatory regime for food. The Sanitary and Phytonsanitary codes of the General Agreement on Tariffs and Trade (GATT) and the Codex Alimentarus appeared irrelevant in the face of rising European resistance to the import of GMOs. However convinced governments might be of the competitive advantages of embracing the new biotechnology, the need to respond to public concerns tended to force them towards moratoria and trade restrictions, which, in turn, provoked the displeasure of the United States government. The national regulatory systems, that had been created in response to concerns over public health and 'novel foods' and the 'release' of transgenic organisms, imposed differing standards in the face of an industry that was operating on a global scale. In some instances as with soya beans, a staple of much processed food, it became impossible to determine whether imports contained genetically modified material or not. More profound issues, especially for the developing countries, were raised by the intellectual property dimensions of biotechnology, of claims to the ownership of the building blocks of life itself, enforced by the new Trade Related Intellectual Property (TRIPs) agreement of the World Trade Organisation (WTO). At the same time spokesmen for the agro-chemical conglomerates, pioneering GMO technology, proclaimed a new 'green revolution' in which their products represented the last best hope of feeding a world population predicted to double by 2046 (from a 1995 total of 5.6 billion) (UNDP Human Development Report, 1998: 177). Much of the campaign against the testing and marketing of GM products was led by the environmental non-governmental organisations (NGOs), such as Greenpeace and Friends of the Earth. They highlighted the potential damage to biodiversity that might be occasioned by the release of transgenic material. Thus, GMOs were added to the list of (potential) international environmental problems that had been constructed during the late 1980s and early 1990s. The first substantive action of the parties to the UN Convention on Biological Diversity (CBD), created in 1992 as part of the Rio process, was to initiate the negotiation of a Protocol on Biosafety, aspiring to regulate the transboundary movement of 'living modified organisms' (LMOs) that might cause environmental harm.

Such were the immediate manifestations of the new biotechnology at the international level. Yet the refinement of genetic technologies clearly had significant implications for an older concern with biological weapons. Might these not now become usable military instruments by virtue of the new technology? How could the incalculable consequences of future genetic warfare be avoided? The other headline-making aspect of biotechnology involved advances in the understanding of human genetics. Some of the implications seemed benign and eminently desirable, but others raised the possibility of interference with what had been thought to be elemental human rights. At the most fundamental level, would it now be possible to engineer human populations, turning some of the wilder eugenicist dreams of the early twentieth century into reality?

It was such questions that prompted a group of IR scholars to come together under the auspices of the British International Studies Association (BISA) during 1998 and early 1999 to consider the implications of the dramatic advances in biotechnological research and applications that were being reported almost daily in the press. Unlike previous technological revolutions in weaponry and communications, the impacts of the new biotechnology on IR, except perhaps in the field of trade politics, were far from self-evident. Contributions were drawn from the various BISA Working

Groups, International Political Economy, Environment, Gender and Security. The representatives of each had a distinctive view of what the new biotechnology meant for them. These views were discussed in two workshops, in Liverpool in October 1998 and Stoke on Trent in May 1999. The results are to be found in the present volume.

To enhance their understanding of biotechnology and ensure technical acuuracy the original participants invited a working molecular biologist, Francis Manning (chapter 2), who teaches in the areas of molecular and cellular biology, to join them. His chapter, which follows this introduction, provides a concise and up-to-date overview of the applications of modern biotechnology and introduces the reader to many key techniques including monoclonal antibody technology and DNA manipulation. The impact of biotechnology is widespread, as befits what is appropriately described as a 'generic technology', capable of affecting many areas of industrial, medical and agricultural activity (Dunning, 1993: 5). Significantly, DNA is itself an information carrier, and modern biotechnology is fundamentally about knowledge. It is effectively an information-based *enabling* technology with DNA and its manipulation (in one fashion or another) at its heart. Knowledge discovered cannot be undone and the long-term consequences of its diffusion are not always clear.

How is the international role and impact of biotechnology to be understood? Technological advance, particularly in weapons and communications, might be seen as driving the transformation of the international system. Nuclear weapons were claimed (under conditions of deterrence) to have revolutionised the conduct of war by severing the Clausewitzian link between political ends and military means. Electronic communications provide the nervous system of a contemporary world system and are the *sine qua non* of globalisation. It has often been claimed that they undermine both the ability of the nation state to control its borders and its very legitimacy. Similar claims have, of course, been made for previous technologies from gunpowder to steam navigation and the electric telegraph. Despite all this there have been few attempts, to date, in the literature of IR, to give technology a central place in explanations of system change (notable exceptions are Skolnikoff, 1993 and Rosenau, 1990). Studies of military technology (and in particular nuclear weapons) apart, there is still little systematic analysis of the process of technology innovation, its diffusion and impact upon IR. Biotechnology is no exception to this rule, although there have been two book-length studies by Russell (1988) and Wiegele (1991) and a detailed comparison between the early US and UK regulatory history (Wright, 1994). On the other hand, rather more has been published on biotechnological developments that impinge on, or overlap with, various subfields in the discipline. Biological weapons provide the primary example, but there is also work on transnational intra-firm collaboration in biotechnology and on trade and environmental/biodiversity issues.

Like much of the economics literature IR has taken technology to be exogenous to its main concerns, a 'given' that indirectly affects the relationships between states. As Gilpin observes:

> Among the so-called exogenous variables that affect the operation of markets are the structure of society, the political framework at the domestic and international levels, *and the existing state of scientific theory and technological development*, all of which constitute constraints and/or opportunities affecting the functioning of economic actors. (Gilpin, 1987: 65, emphasis added)

A trend is emerging that seeks ways to place technology and knowledge more centrally into international political economy (IPE) and IR (Talalay, Farrands and Tooze, eds, 1997; Strange, 1988; Jones, 1995) and, in keeping with the innovation economists (Foray and Freeman, 1993; Archibugi and Michie, 1997), in a way that denies that technological innovation can be relegated to the 'exogenous' realm. Indeed innovation economists offer concepts useful in explaining the innovation process and the diffusion of technology, as well as offering a halfway house between the IPE and international business literatures. The chapters by Pownall (chapter 7) and Russell (chapter 6) engage these perspectives – and others – but add a political context. While the current book, in part, represents a contribution to internalising technological concerns within IR/IPE, it was clear from the outset that biotechnology could not necessarily be treated in exactly the same way as other technologies, and particularly those that have been associated with revolutionary change in the international system.

Another technological revolution?

At first sight there are some evident similarities between biotechnology and other technologies that have had a profound impact upon international relations – notably nuclear weapons and communications. All have been viewed deterministically. In this view, if something can be done, it will be done, and the exploitation and application of scientific advance has an unstoppable dynamic. Once cloning techniques are developed to the point at which human beings can be engineered, they will be applied regardless. Once a genetically discriminating weapon is feasible, it will be developed. Such a fatalistic surrender of control is both morally and empirically unacceptable. Technological change cannot be realistically regarded as a deterministic force in world affairs, the relationships between knowledge, political and socio-economic structures and even ideology are far too complex and subtle to allow us to set up technological change as some kind of master independent variable (Nau, 1974; Vogler, 1981). Yet there is still a sense of what Winner (1977) has called 'reverse adaptation', where the process of technological change acquires a certain autonomy simply because it is believed to provide the key to future power and prosperity. While there are differences in the estimation of market size for biotechnology products and in expectations of its growth, differences in business willingness to invest in technology, differences in the willingness of national funding bodies to promote the transfer of biotechnological knowledge to industry, and reluctance amongst even the most ardent supporters to see this knowledge as a panacea, no industrialised country would wish to fall too far behind or lose touch with its progress. This has as much to do with maintaining a place at the leading edge of advanced technology as the achievement of specific commercial returns.

One of the more unhelpful conclusions of much of the older literature on technology and IR is that the consequences of change are inherently 'double-edged' and even contradictory (Falk, 1975; Kinter and Sichermann, 1975). Thus new communications technology has the effect of both undermining the state, by facilitating message flows across borders, while simultaneously yielding quantum increases in the capability of the central authorities to eavesdrop on their citizens. Biotechnology can be portrayed in a similar way. The planting of genetically engineered crops provides a typical example. On the one hand, there are potential environmental benefits in the reduction of pesticide use that such crops are claimed to allow. On the other, there are a range of often unanticipated side effects involving, for example, the destruction of butterflies and the potential for ecological disruption occasioned by the introduction of alien engineered and self-regenerating organisms.

Despite all the potential benefits, the reaction to the products of the new biological sciences is similar to that which met the development of nuclear fission in at least one important respect. There is a profound and visceral fear of 'men who play god' (Moss, 1970). Technology has always involved rearrangement of the natural order, to the extent that after millennia of human intervention the very concept of a 'natural order' is difficult to sustain. Yet nuclear and now biotechnology seem qualitatively different to the extent that they involve human manipulation of the fundamentals of existence. In the case of biotechnology a re-engineering of life itself. As the chapters by Dyer (chapter 3) and Bretherton and Stephenson (chapter 4) illustrate, this must raise the most profound ethical questions with wide international ramifications. The anxiety about GM foods reflects deeply embedded cultural understandings. Not only do cultural norms allocate to women primary responsibility for their families food, they also determine what may and may not be eaten. Thus the strong distaste expressed by women for 'unnatural' GM foods reflects not only their concern for family health, but also their responsibility (actual or residual) for ensuring that cultural norms concerning food purity are properly respected. Ultimately the establishment and maintenance of cultural norms and rules, in relation to food as elsewhere, reflects the distribution of power within society. In many cultures food has been accorded great significance. So too have fertility, birth and death. Modern biotechnology is deeply disturbing because it challenges established meanings in these fundamental areas. Just as we are no longer sure what our food is made of so the basic categories of human life and death may be changing their meaning.

Many technologies are characterised, as is modern biotechnology, by rapid growth and uncertainty. What sets biotechnology apart from the rest, however, is that living organisms are not only modified but can then replicate themselves autonomously. This adds a new dimension of unpredictability and substantially complicates the application of the 'precautionary principle' to innovations. As yet much of the debate about the ecological and other impacts of living modified organisms is essentially speculative, because there have been relatively few 'releases'. GM technology tends to confound established 'risk assessment' procedures which require that both probabilities and outcomes be relatively well defined. Instead this new technology is characterised not so much by uncertainty but typically by ignorance (ESRC, 1999: 6). The processes involved do not have the kind of predictability of, for example, chemical reactions which can serve as a basis for regulation. This may entail the requirement for case-by-case assessment, although the EU and US regulatory architecture differs widely. Debates over genetic engineering have occurred within the national jurisdictions of many developed and, more recently, developing states; although the concerns and priorities have differed widely. In Europe national differences created particular regulatory difficulties for the EU which were a precursor of wider international problems. In the United States the debate about GMOs was widely regarded as having been resolved some time ago, leaving Monsanto and other corporations to press ahead with substantial commercial application of the new technology. The nagging anxiety must remain that there is a danger of creating an unforseen problem equivalent in kind to that of nuclear waste but possibly on a more all-pervasive scale.

The organisation, funding and processes of biotechnological research differs greatly from the model of the post-war nuclear industry, dominated by the military industrial complex or even from the telecoms and computing industries with their massive state-supported research and development (R&D). The actual conduct of biomolecular research can be small scale and does not require a great national effort with a huge infrastructure. The key players are firms and research teams. The major corporations often use smaller research companies – sometimes described as research boutiques. Large firms around the world have for example, established various arrangements with the new biotechnology firms amassed in the United States. The big company avoids the difficult choices of where to develop in-house biotechnology expertise with the attendant risk of missing the winners. The smaller specialist firms find their financial life-blood boosted and on occasion the academics with the foresight to build a company around a promising idea become paper millionaires. A critical difference in comparison with the nuclear or even the telecoms industry is that civil applications lead the way. Whereas the former (nuclear power generation, civil use of space, the internet) were 'spin-offs' of military R&D, the miltary applications of biotechnology are 'spin-ons' of essentially commercial and civil developments.

The commercial exploitation of biotechnology leads to some novel issues in relation to ownership. The building blocks of life are being 'commodified'. Here, the debates over GMOs have a new North–South dimension. Whereas the traditional demand of developing countries was to ascribe common heritage status to resources, such as seabed minerals or GSO orbital positions, the spatial location of the bulk of untapped biodiversity/genetic resources in the South has meant that they have an interest in asserting national rights. The Convention on Biological Diversity became something of a battleground between the vested interests of biotechnology firms – sponsored by Northern governments – and states of the South waging a rearguard action to belatedly re-establish some semblance of control over their natural resources. The firms were keen to ensure unfettered access to the unexploited mass of genetic diversity contained in the rainforests and other territories of the South. This clash of interests spilled over into the negotiations leading to TRIPs, as the protection of intellectual property subsumed debates over patenting GMOs. How much would a plant need to be altered by genetic manipulation techniques to be 'novel' and therefore protected? How extensive would that protection be around the world? And what if that plant came from the unexploited lush areas of the South? In reaching these international agreements as ever it has been found that the 'devil is in the detail' – the consequences of which are explored by Vogler and McGraw (chapter 8), Williams (chapter 5) and Falkner (chapter 9).

A final and most politically significant feature of the new biotechnology concerns the distinctive character of the relevant 'epistemic communities' – those knowledge-based transnational networks of specialists, which Haas (1992) and others have seen as possessing critical influence in the interplay between science and policy. Here, there are real differences in the wider scientific community between the molecular biologists and the ecologists. Their contrasting paradigms may be described in terms of atomism and holism. Despite attempts at public reassurance through presenting a 'scientific view' of minimal risk of say GM food there is scientific contestation. In the UK the policy community has been slow to appreciate the special difficulties of managing the new biotechnology and the need to understand the 'framing assumptions' which shape and often narrow the scope of supposedly 'objective scientific assessment'. Indeed, 'faced with political crises surrounding new issues such as GMOs there appears to be less enthusiasm for admitting the limits to relevant scientific knowledge' (ESRC, 1999: 7) Attempts at public reassurance, by reference to scientific knowledge, or perhaps the absence of any scientific knowledge, of the harmful effects of GMOs, have tended to be counterproductive in the face of increasing public scepticism. Such recent episodes have serious implications for the public authority of science and for the ability of governments to pursue 'rational' policies in the face of public distrust and anxiety. This may be a passing concern in Europe, or it may represent something much more profound. It also contrasts with earlier periods when, although the ethics of pursuing nuclear technology at the beginning of the cold war were fiercely debated, the authority of the science itself remained inviolate.

The book

During the course of the two workshops giving rise to this volume we wrestled with the problem of the order in which the various contributions should be presented. We might have commenced with a mainstream IR analysis. Two reasons led us to reject this path. First, we do believe that the reader – as we did – will benefit from the insights of Manning (chapter 2), a professional scientist, into the development, range and application of the technology. His chapter follows this introduction. Second, although most of the contributors might identify themselves as political scientists, interested in aspects of IR, the book should engage a wider audience. Biotechnology is itself pervasive and many disciplines and fields of study are interested in understanding its progress and impact. Thus we open with Manning's review of the technology followed by Dyer's (chapter 3) assessment of the broader normative implications of biotechnology in an IR setting. The first section is completed with Bretherton and Stevenson's (chapter 4) provocative commentary on society's embrace with the techno-scientific revolution, as witnessed from a women's perspective.

Our workshops, under the auspices of BISA, involved members of BISA working groups. Two of these were formally concerned with IPE and the environment and IR, respectively. In our discussions we found it difficult to disentangle the IPE and environment concerns when the subject was new biotechnology. Is biodiversity an environment issue or an IPE issue? Likewise is GM food one or the other? In short we concluded that a more cohesive approach was to focus part II on these joint concerns. Thus, Williams (chapter 5) reviews the politics behind the establishment of the world trade regime provisions for the protection of intellectual property, negotiated under the GATT, operated by the WTO, and with huge significance for biotechnology. Following on from Williams' insights on the pressure firms used to achieve a favourable TRIPs outcome, Russell (chapter 6) considers how biotechnology, as a generic technology, might be conceptualised in IPE. His suggestions draw upon Actor-Network Theory (ANT) and the work of innovation economists. Pownall (chapter 7) then explores the efforts of regions to bolster the activities of their biotechnology industries in a competitive global political economy. The attempt to establish an environmental regime for the new biotechnology is then considered in the chapter by

Vogler and McGraw (chapter 8). Their consideration of the difficulties encountered in negotiating the Cartagena Protocol on Biosafety illustrates the tensions between trade, development and environmental objectives. Such complex problems are the focus of the concluding chapter in the section by Falkner (chapter 9) who surveys international trade conflicts and agricultural biotechnology.

The final section brings together four aspects of one of the main concerns of contemporary security studies – biological warfare. Dando (chapter 10) opens the section with an overview of the 'classical' agents underpinning the development of biological weapons and their feared use in war. Biotechnology can profoundly shape future developments of such weapons, from 'designer genes' to 'designer weapons'. However, as Whitby (chapter 11) reminds us, not all biological warfare is to do with the deployment of weapons against people. A very serious concern is the past and future use of weapons targeting crops, including those that are the staple diets of large sections of the world's population. Having indicated the potential for various forms of biological warfare, the final section, in Littlewood's contribution (chapter 12), includes an assessment of the international progress towards effective biological weapons control. If controls fail the state-system may be left with little else other than the responses catalogued by Spear (chapter 13), which are founded on more traditional security and deterrence principles. The final chapter attempts a synthesis of these diverse, yet interconnected, strands in order to arrive at some preliminary conclusions on the implications of the new biotechnology for the contemporary study of international relations.

Part I

Science, ethics and gender

In Part I we consider the scientific character of biotechnology, its origins and applications and some of the key normative issues that arise. Manning (chapter 2) provides a necessary basis for the rest of the book. His chapter provides us with some basic definitions of biotechnology along with an outline of its historical development, emphasising the distinctive features of the 'new' biotechnology. Using largely non-technical language Manning describes the significant applications of biotechnology which form the substance of debate in succeeding chapters. He concludes with a brief discussion, from a practitioner's viewpoint, of some of the risk assessment issues that, as we have already seen, are so especially troubling for our attempts to come to terms with this new technology.

Perhaps more than any other type of innovation the new biotechnology has raised ethical and normative issues. For a technology that seeks to manipulate the elemental building blocks of life this could not be otherwise. However, the normative discourses of IR do not fit comfortably with the values associated with biotechnology. Dyer (chapter 3) analyses the tensions and contradictions. Amongst these are the conflict between anthropocentrism and ecocentrism, the globalised character of the biotech industry and the values and assumptions of the inter-state system and specificity or universality of knowledge and conceptions of justice. He concludes that the IR of biotechnology 'are and will continue to be, conditioned by emerging values and value structures which are largely unfamiliar to the traditions of IR'.

Such traditions have privileged states rather than people and frequently involve masculinised discourses on power, knowledge and security. As Bretherton and Stevenson (chapter 4) demonstrate in the final chapter of the section, a consideration of the new biotechnology can bring such, often inarticulate, assumptions into sharp focus. There is a gender dimension to most of the discussion in this book. It is particularly marked in the political economy of development but the authors focus upon the particular anxieties and

ambivalence that arise for women when confronting the life-creating potential of biotechnology – women having been stereotypically associated with
birth and nature. Biotechnology promises substantial benefits but also a
usurpation of the traditional female role. In the starkest terms it forces us to
address the question of the ethical acceptability of technologically driven
'progress'. Ultimately this question lurks behind all the discussion of industrial structure, trade, environmental protection and security.

2

Biotechnology:
a scientific perspective

Francis C. R. Manning

The nineteenth century was very much the 'Age of Chemistry' when huge strides were made in the understanding of chemical processes. Similarly, the twentieth century could be regarded as the 'Age of Physics' with the advent of fields such as general relativity, quantum mechanics and atomic energy. If present trends continue, biology is likely to become the pre-eminent science of the twenty-first century. Biotechnology will increasingly affect people's lives, both directly and indirectly.

The aim of this chapter is to provide an introduction to the area of biotechnology for the non-specialist. Biotechnology differs from other types of technology in that it deals with living organisms that have the potential to reproduce themselves autonomously in the environment without human intervention. This produces the potential for new benefits but also new types of risk. The environmental release of a toxic chemical, or even a radioactive isotope, can have only finite consequences because only a limited amount of material is released. In contrast, once released, a living organism has the potential to replicate itself and, therefore, the effects of such a release may increase rather than diminish over time. In the case of genetically modified organisms there is also the potential for unforeseen interactions with other biological entities in the environment.

The term biotechnology was coined almost eighty years ago by the Hungarian Engineer, Karl Ereky (Murphy and Perrella, 1993a). It was originally defined as the means by which products could be made from raw materials using biological processes. In this sense biotechnology has been used by human beings for many centuries in the form of selective breeding of animals and plants, and the use of micro-organisms to produce commodities, such as wine and cheese. However, since the 1970s, human ability to manipulate biology has increased dramatically. In particular, the advent of recombinant DNA technology has allowed the manipulation of the genetic material of organisms to create entities that are not found naturally.

What is biotechnology?

Many definitions of biotechnology exist; however, they are often too general or specific to be useful. Possible definitions include:

- 'The production of products from raw materials with the aid of living organisms' (Karl Ereky, 1919)
- Application of scientific and engineering principles to the processing of materials by biological agents to provide goods and services
- The industrial use of living organisms or biological techniques *developed* through basic research
- The technological use, through science or engineering, of living organisms or *parts* of living organisms, in their natural or *modified* forms.

Ereky's original 1919 definition can be stated as: 'Any process where a biological organism is used to make a product for human use.' This definition would include processes such as traditional fermentation technologies that produce alcohol and cheese-making. It is therefore probably too broad to be a useful definition of modern biotechnology. Another problem with this definition is that it does not recognise the increasing use of modified biological entities. At the other extreme, in the popular imagination, biotechnology has become synonymous with genetic engineering. In reality, biotechnology is a highly interdisciplinary 'hybrid' field (Smith, 1996). This can be illustrated by examining the composition of a biotechnology project team. Modern biotechnology projects use expertise from numerous disciplines that can include medicine, engineering, biology, chemistry, agriculture and, increasingly, bioinformatics (Bains, 1996, Lee, Chin and Mosser, 1998). Within each discipline there can often be several specialists with differing skills. So, for instance, biologists may be specialists in molecular biology, cell biology, biochemistry, microbiology, immunology, protein engineering and other areas. A useful working definition of biotechnology is 'the technological use, through science and engineering, of living organisms or parts of living organisms in their natural or modified forms'. This definition recognises several of the important points: that biotechnology is multidisciplinary and draws on many science and technological disciplines; that both whole organisms and parts of organisms are used; and that increasingly the biological entities employed may be modified through the use of science and technology. It is likely that definitions will continue to evolve in parallel with the technology available.

Biotechnology in historical perspective

To understand why biotechnology is becoming a major influence at this time it is useful to review briefly the advances that have led to the present position. Some of the major milestones are as follows:

14

<1750   Selective breeding of plants and animals, brewing, cheese
         production, baking
>1750   Crop rotation
1797   Edward Jenner's development of a smallpox vaccine
1802–69 The term 'biology' used. Proteins, enzymes and DNA isolated
1893   Pasteur awarded first patent for a microbial fermentation process
1914–18 German fermentation technology
1919   The term 'biotechnology' coined
1928   Discovery of antibiotics
1944   Penicillin produced in bulk
1953   Structure of DNA elucidated
1973   First recombinant organism produced
1975   Monoclonal antibodies
1977   Human growth hormone made by bacteria
1978   Genentech Inc. produce human insulin in bacteria
         First test tube baby
1980   Patenting of genetically modified microbes
1983   First genetically modified plants
1984   DNA fingerprinting
1985   Patenting of genetically modified plants
1988   Patenting of genetically modified mammals
1993   Flavr Savr genetically modified tomatoes sold in US
1997   Cloning of 'Dolly the Sheep'

The technological use of biology by humans dates back at least 4000 years, the use being of yeast to ferment beer known to the ancient Sumerian's 2000 BC. However, around the eighteenth century only a few basic biotechnological processes were practised, such as the selective breeding of plants and animals, the use of yeast for brewing alcohol and baking and the use of bacteria in cheese production. In Europe crop rotation began to be used to increase productivity, but the first clearly recognisable application of modern biotechnology in the Western world was probably the use of cowpox to immunise humans against small pox by Edward Jenner in 1797. This approach of using an attenuated (less harmful) organism to immunise humans against a dangerous organism is still practised today. One potential current application is in the development of a vaccine against the human immuno-deficiency viruses (HIV) that cause Acquired Immuno-deficiency Syndrome (AIDS) (Boily *et al.*, 1999).

The early nineteenth century saw the term 'biology' first used, as the study of biological systems began to grow into an established science (Murphy and Perrella, 1993a, 1993b). By 1869 DNA and proteins were recognised as important cellular components, although little was known about their function. Modern biotechnology began to emerge in 1893 when Pasteur was granted a patent for a microbial fermentation process. This landmark effectively paved the way for the commercial exploitation of novel biological organisms that is continuing to grow today. Biotechnology continued to grow during the early twentieth century, in large part as a result of military requirements. During The First World War the Germans, in particular, used biotechnology in the form of novel microbial fermentation processes to produce raw materials for the production of explosives for the war effort (Murphy and Perrella, 1993a, 1993b). In the 1920s Alexander Fleming discovered the first antibiotic, penicillin G but it was not until World War II that penicillin was produced in large quantities (Hare, 1970).

DNA (deoxyribonucleic acid) is the molecular blueprint for life and codes for the proteins that perform the functions of cells. In 1953 James Watson and Francis Crick published their structure of DNA and effectively began the science of molecular genetics (Watson and Stent, 1998). DNA consists of a series of molecules called bases that join together to form a linear strand. DNA contains four types of bases termed A, T, G and C. The order in which these bases occur on the DNA strand determines what information is carried by that strand. This information is divided into regions that are called genes. Each gene codes for a specific protein. The technique by which the order of bases on a DNA strand is determined is called DNA sequencing. This technique allows molecular biologists to 'decode' the information held in an organism's DNA.

In 1973 the first recombinant (genetically modified) organism, an *Escherichia coli* (*E.coli*) bacterium, was produced (Murphy and Perrella, 1993a, 1993b). A notable development occurred in 1975 when monoclonal antibody technology was developed (Köhler and Milstein, 1975). Patents on this technique led to a huge international business that has spawned numerous commercial products (Cambrosio and Keating, 1995). By the late 1970s, both human growth hormone and human insulin had been produced in bacteria and in 1980 the first patent for a genetically modified micro-organism was granted in the US (Gibert School, 1999; Murphy and Perrella, 1993a, 1993b). This was rapidly followed by the development of genetically modified plants and their patenting in 1985. The first genetically modified animal was patented in 1988. The number of patents granted for genetically modified organisms, and in particular plants, is increasing rapidly.

Two factors that have been, and will continue to be, important in the development of biotechnology are necessity, in the form or medical or other need such as war, and the ability to gain economic benefit from advances through patents.

Current applications of biotechnology

Biotechnology is a diverse undertaking that will increasingly affect people's lives. Notable fields of special interest are medicine, agriculture, the environment, industry and forensic science. Some of the most important areas of these fields are:

Medicine
- Monoclonal antibody-based tests
- Polymerase chain reaction-based tests
- Recombinant proteins used as drugs

Agriculture
- Modification of animals by transgenic technology
- Modification of animals by recombinant drugs/antibiotics
- Crop improvement/modification by transgenic technology

Forensics
- DNA fingerprinting

Environmental/industrial decontamination
- Microbial detoxification of waste
- Environmental monitoring

Medicine

Biotechnology is increasingly used in medicine in the area of diagnostics and also for the production of drugs, such as insulin (Howey *et al.*, 1982) and vaccines (DelGiudice, Pizza and Rappuoli, 1998). Because of the importance of these areas they will be briefly reviewed.

Monoclonal antibodies

One of the earliest examples of the potential economic benefits of biotechnology was the development of monoclonal antibody technology in the mid 1970s (Köhler and Milstein, 1975). This technology is central to many medical diagnostic procedures (Cambrosio and Keating, 1995). Usually when an animal's immune system is exposed to a foreign entity (an antigen), such as a new protein, immune cells called B-cells make many types of antibodies that each bind to different parts of the antigen. These 'polyclonal antibodies' have been used extensively but they have one major drawback, they can only be made in small amounts by repeatedly injecting an animal, such as a rabbit, goat or sheep, with the antigen. The polyclonal antibody is then made from the animal's blood. If more antibody is needed a new animal has to be immunised and the antibody produced may have different chemical properties to the original. In contrast, monoclonal antibodies are made from immune cells that have been bio-engineered to produce only one type of antibody molecule (Genentech, 1998). These cells can be grown in the laboratory and allow an almost limitless supply of a specific antibody to be produced. The specificity of these antibodies also allows very small amounts of a biological agent to be detected and accurately quantified. Monoclonal antibody-based procedures have revolutionised diagnostic medicine and numerous hospital laboratory tests use them. They are even been used in over the counter health products, such as home pregnancy tests that detect the increase in human chorionic gonadotrophin (HCG) hormone that occurs in women's urine early in pregnancy. Despite the fact that monoclonal antibodies were developed at the Medical Research Council laboratories in Cambridge, England, the economic benefits of this technology were largely realised by a United States-based company that patented this process. Hence, the traditional means of 'staking a claim' on scientific discovery through publication of results is not sufficient to quarantee commercial rights on that research. It can be argued that this may cause scientists to delay publication of important findings in some instances until patents have been obtained. This is probably more common in private/industrial than public/university laboratories.

Production of recombinant proteins by genetic modification of cells

Recombinant DNA technology is the name given to the techniques that are used to manipulate DNA in the laboratory. One of the uses of this technology is in the creation of genetically modified organisms (GMOs). A general scheme for the creation of a genetically modified organism is illustrated in figure 2.1. This procedure has four basic steps (Old and Primrose, 1994). First the DNA (gene) that codes for the protein of interested is ligated (joined) to an expression vector to produce what is termed a recombinant DNA molecule. The expression vector is a piece of DNA that contains sequences that allow the gene of interest to be inserted into the DNA of a host cell and translated into a protein. The second step is introduction (transfection) of copies of the recombinant DNA molecule into a host cell. This cell may be a bacteria, yeast, cultured mammalian cell, plant cell or a fertilised mammalian egg. At the present time the methods used to insert DNA into host cells are not completely effective and some cells do not receive a recombinant molecule. In the third step of the procedure it is therefore necessary to select host cells that have taken up a recombinant DNA molecule. This selection is achieved by including a gene that acts as a 'selectable marker' in the design of the expression vector. Usually this is a gene that encodes resistance to an antibiotic drug, such as kanamycin. The final step is to propagate the genetically modified cells, or organisms, to create a clonal population within which all members are genetically identical.

Bacteria such as *E.coli* have been genetically engineered since the earliest days of recombinant DNA technology (Primrose, 1987; Old and Primrose, 1994). Compared to mammalian or human cells, bacteria have very small genomes that are relatively simple to alter. It has been possible to insert a human gene into bacteria so that it makes a human protein for decades. Several simple proteins, such as insulin, have been produced by such genetically modified bacteria (Howey *et al.*, 1982). One major advantage of this technique is that bacteria are cheap to grow on a large scale. However, this

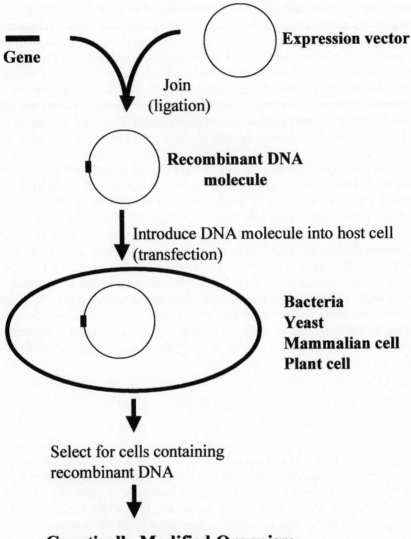

Figure 2.1 Production of recombinant proteins by genetic modification of cells

approach has limitations. Proteins from higher organisms often require folding in very specific ways, or require the addition of complex sugar molecules, to have the correct biological activity (Jenkins, Pareth and James, 1996). Much of the biological machinery needed to make this happen is not present in bacterial cells. Therefore bacteria cannot always process a human protein in the way it would normally be processed.

To try and circumvent problems with protein processing, yeasts have been used to make recombinant proteins instead of bacteria (Gellissen and Hollenberg, 1997). Yeasts, like humans, are eukaryotic. Their cells have nuclei in which their genome is contained and they are much closer to us in evolutionary terms. However, their biology is obviously not identical to ours and they too may not always process a protein as a human cell would. They do, however, offer the potential of producing large amounts of a recombinant protein relatively cheaply. In addition, a great deal of research into the genetic manipulation of yeasts has been done and strategies to express successfully a foreign protein in these cells have been available for some time.

From a biological perspective, the best type of cell for producing a human protein that is correctly processed and folded is a human cell. Many types of human cells can be grown in tissue culture in the laboratory (Mather and Robert, 1998). Often such cell lines were originally derived from human tumours because, unlike most normal human cells, cancer cells have an unlimited lifespan in culture. In recent years, our abilities to understand and manipulate the genetics of animal and human cells have increased considerably (Lewin, Siliciano and Klotz, 1997). It is now quite possible to express a protein in cultured mammalian cells, although the technologies involved are far from perfected. Proteins produced in this way tend to be expensive owing to the high cost of growing cells in culture. The major commercial application of this technique to date is to produce high-grade, high-cost proteins for medical use (Rohricht, 1999).

Agriculture

Biotechnological modification of animals

Humans have selectively improved domesticated animal species for many centuries. The power of this approach can be easily seen just by considering the numbers of varieties of dogs in the world today that have arisen from selective breeding from a common ancestor (American Kennel Club, 1997). Biotechnology is increasingly giving us the ability to make highly selective changes in animals. This may arise either through direct alteration of specific genes, or through hormonal or pharmaceutical interventions. Perhaps the most contentious procedure has been the development of transgenic (genetically altered) animals that have had specific genes added or removed from their DNA (Pinkert, 1994). This technology is still not completely reliable and numerous attempts often have to be made to produce a transgenic animal. Nonetheless, transgenic animals (usually mice) are a very valuable research tool that contribute to our understanding of many diseases, including cancer (Anderson, 1988) and heart disease (Christensen, Wang and Chien, 1997). It needs to be mentioned that transgenic experiments are expensive in both time and money. They are therefore not undertaken lightly by research scientists.

A commercial application of transgenic technology is to produce a valuable protein in a domesticated animal (Di Berardino, 1999; Loi *et al.*, 1997; Schnieke *et al.*, 1997; Wilmut *et. al.*, 1997a; Wilmut, McWhir and Campbell, 1997b). Several animals have already been genetically modified to produce a human protein in their milk. One such case is sheep expressing the human blood clotting Factor IX protein. This protein is needed for the treatment of some haemophiliacs who do not make Factor IX themselves. Potentially a small flock of transgenic sheep could supply the global requirements for this protein. Unfortunately, the offspring produced by such animals from normal reproduction often do not express the transplanted gene and therefore do not produce the wanted protein. Since it takes considerable time and money to produce such a transgenic animal, there is a great economic incentive to find ways to reproduce it without losing the activity of the transplanted gene. One potential way would be to clone such an animal. Since the clones would be genetically identical to the parent they could reasonably be expected to express a transplanted gene as well as the parent. This was the original reason for cloning 'Dolly the Sheep'. At the present time the low success rate of cloning (about 2 per cent) and high cost means that it is not an economically viable option for producing normal farm animals. Based on the animal experiments so far conducted, any attempt at human cloning would currently require fifty women who were willing to act as surrogate mothers for implanted embryonic clones to generate one cloned child. At least some of these women would be likely to give birth to a deformed or handicapped child. In the light of these technological requirements alone, human cloning seems unlikely to become widespread until the efficiency of this procedure is dramatically increased.

Transgenic technology allows very specific changes to be made to the DNA of animals, however, this has to be viewed in perspective. At the present time only one, or possibly a few, genetic changes can be made in an animal. Assuming the genome of a mammal contains about 100,000 genes (Uddhav and Ketan, 1998), adding a new gene would change that animal at the genetic level by only 0.001 per cent. In comparison selective breeding techniques can change a species by up to 0.1 per cent per generation. Therefore, although molecular biology allows very specific and possibly novel gene changes to be made, its absolute power to change an animal's DNA is far less impressive. Put another way, anyone wishing to create a 'master race' of humans would at the present time be far better served by selective breeding than by advanced molecular genetics.

To date, probably the biggest impact of biotechnology on farming has been in the form of drugs aimed at improving the health or productivity of animals. The extent of drug use by farmers has increased greatly, particularly in the United States, where growth enhancing agents, such as somatotropin, are commonly given to animals (Etherton and Bauman, 1998). Clearly the use of such drugs provide economic benefits. Scientific opinion is, however, still divided on the health risks of these practices. Concerns about introducing growth-promoting agents into the human food chain stem from the observation that some cancers have been linked to exposure to growth-promoting agents. For instance, the risk of a woman developing breast cancer is partly related to how long her body is exposed to oestrogen, the hormone that promotes breast growth (Cauley *et al.*, 1999; Wu, Pike and Stram, 1999). In the case of antibiotics several potential problems have been suggested (Adams and Templeton, 1998; Jensen, 1998). One of the most worrying is the possibility that the increased levels of antibiotics entering the human food chain may increase the rate at which some bacteria are becoming resistant to these drugs. At the present time, the ever-increasing numbers of bacteria that are completely resistant to antibiotics is one of the major global health concerns.

Plant biotechnology

Biotechnology, in the form of genetic modification, is currently being applied to agricultural plants in order to introduce highly specific changes that were not possible using traditional selective breeding. Potentially these changes could improve crop yields, food quality, nutritional content, pest resistance, herbicide resistance or produce new materials (Oksman, Caldentey and Hiltunen, 1996; Panopoulos, Hatziloukas and Afrenda, 1996; Pauls, 1995; Ranalli and Cubero, 1997). This technology may also have benefits for forestry (Walter *et al.*, 1998) and plant conservation (Harding, Benson and Clacher, 1997). To date, however, the traits added to plants have been mainly related to profit rather than improving the plant for nutritional or other reasons. Most notably has been the production of a soya plant resistant to glycosate-based fertilizers by Monsanto (Hart and Wax, 1999).

At least two types of problems with the use of genetically modified food plants have been suggested. The first type is related to the effects that genes added to these plants may have when consumed by humans and centres around questions as to whether a gene, either directly or indirectly, may encode a protein harmful to a human (Union of Concerned Scientists, 1999). Some concern has been voiced regarding the unavoidable use of antibiotic resistance genes, such as the kanamycin gene, in the process of creating transgenic plants. It has been suggested that these genes might be passed on to micro-organisms in the intestines of humans consuming the plant. Current opinion suggests that the risks from this are negligible (Ruibal, Mendieta and Lints 1998); however, this may not always be the case. One area of particular concern is whether a gene might encode a protein that causes an allergic response (Union of Concerned Scientists, 1999). At the present time, the knowledge base from which to make accurate risk assessments of this type is still growing, and caution obviously needs to be used at each stage.

A second concern stems from what has been termed 'genetic pollution'

(Parker and Kareiva, 1996). To what extent will a genetically engineered plant spread into the environment either alone or by cross breeding with related plants in the environment? What would the effects of this be on the ecosystem? Informed speculations can be made as to the answers to these questions. It has to be noted, however, that science has been relatively weak at predicting potential outcomes when a foreign species is introduced into a new environment. Therefore it is probably prudent at the present time to proceed cautiously with the release of genetically modified plants into the environment. Experimentation will be needed to develop a knowledge base from which risk assessments can be made.

Biotechnology and forensics

The technique of DNA fingerprinting has become a mainstream approach in the area of forensic science as well as being widely used for paternity testing (Jeffreys, Wilson and Thein, 1985; Gill, Jeffreys and Werrett, 1985). This technique essentially detects the pattern of a type of DNA sequences called mini-satellite sequences in an individuals DNA. Each of us inherits a unique pattern of mini-satellite sequences from our parents. By analysing these sequences in cells left at a crime scene, they can be linked to an individual. In theory, virtually no one on the planet would share the same pattern of these DNA sequences. There is, however, a potential problem with this technique. In order to analyse small amounts of DNA it is commonly amplified many million times by the polymerase chain reaction before being analysed by DNA fingerprinting (Mullis, Ferre and Gibbs, 1994). The power of this amplification means that tiny levels of contaminating DNA can potentially give false positive results unless the procedure is performed under strict conditions. Given this potential for widespread error, DNA fingerprinting of the general population would be highly worrying. In addition to this are the civil liberty issues connected with any government having a genetic data base of its citizens.

Future applications of biotechnology

Predicting future directions in any scientific discipline is fraught with difficulties and this is particularly true of the biological sciences. Many of the most influential advances have been unexpected and come from basic rather than applied or 'directed' research. An excellent example of this is the development of the polymerase chain reaction (PCR) which has revolutionised molecular biology (Mullis, Ferre and Gibbs, 1994). The PCR method uses a novel enzyme called *Taq* DNA polymerase to chemically copy a specific piece of DNA several million times. This enzyme was originally obtained from the bacterium, *Thermophillus aquaticus*, that was discovered as a result of basic research on the microbes that live in hot springs. In the early 1980s, when PCR was developed, it is doubtful that many working scientists could have predicted that such an apparently esoteric field would have yielded a multi-million dollar industry within a few years. Nonetheless, by extrapolating current trends some predictions can be made with varying degrees of confidence. Some likely future applications of biotechnology are listed below.

- Completion of the Human Genome Project
- Elucidation of gene interactions
- Widespread genetic typing of individuals
- Pre-implantation diagnosis of human embryos
- Behavioural genetics
- DNA fingerprinting of the general population
- Gene therapy
- 'Designer' tissues for transplantation
- Designer babies
- Human cloning

It should be noted that the speed at which this technology is advancing means that the boundary between current and future uses of biotechnology is necessarily blurred.

The human genome project and elucidation of gene interaction

The human genome project is the name given to the global attempt to identify and determine the sequence of every gene present in the human genome (Uddhav and Ketan, 1998). Given the rate of progress of this project it can be predicted with reasonable certainty that the entire human genome will be mapped before 2001. It is hoped that this will allow the identification of all the genes involved in human disease. Initially, disorders involving one or only a very few genes will be the first completely characterised by this research. These diseases are, however, relatively rare in the population (Morgan and Anderson, 1993). By far the major causes of death in the Western world are the multifactoral diseases, such as cancer and heart disease. The risk of developing these diseases appears to be influenced by numerous environmental factors as well as some of the genes that a person inherits. To what extent individuals' genetics affect their risk of developing cancer or heart disease is currently unknown. This type of genetic risk analysis will require a detailed understanding of how genes interact both with each other and with environmental factors.

Widespread genetic testing

If the human genome project leads to the characterisation of most, if not all, of the defective genes that can be inherited by humans, it could be argued that widespread genetic testing of the population could have major health benefits. Individuals who have inherited genes that put them in high risk categories for major killers, such as coronary heart disease and cancer, could be identified and put on preventive regimes. At the present time genetic testing is a relatively expensive and time-consuming process and is therefore reserved for individuals with a family history of an inherited disease. The recent development of Microarray DNA testing technology will almost certainly change the current approach to genetic testing (Lemieux, Aharoni and Schena, 1998). Within a few years it will be possible to test an individual and know exactly what defective genes they have inherited. This may cause as many problems as it may solve. The most beneficial use of this technology would be where an individual at risk of contracting a treatable disease is identified early. Let us consider the case of a woman who has inherited one of the breast cancer-linked genes BRCA-1 or BRCA-2 (Mark and McGowan, 1996). Such a woman would have a very high risk of developing breast cancer, up to 90 per cent in some studies. It could be argued that this knowledge would be beneficial because this individual could be regularly screened and any cancer would be more likely to be detected at an early stage when treatment is most effective. It is clear, however, that some women would prefer to take the radical step of having a double mastectomy rather than live with the worry of developing breast cancer. Neither of these options is a particularly attractive one but at least in the case of cancer there are medical interventions available. One of the under-appreciated outcomes of modern cancer research, however, is that while cancer genetics has increased our ability to determine if an individual is at high risk of developing some cancers, it has not yet yielded new treatments that can be given to most cancer sufferers. While a miracle cure for cancer may never be developed there is, however, likely to be continued incremental improvement in treatments and cure rates for most cancers.

The potential ethical dilemmas of genetic testing become clearer in the case of an untreatable disease such as Huntington's chorea (Furtado and Suchowerksky, 1995). Individuals suffering from this genetic disease suffer a prolonged period of physical and mental deterioration in their middle age that inevitably results in death. Although a genetic test for this disease is available, there is no medical intervention to help the sufferer. Therefore, there is no direct health benefit to the affected individual in performing the test. Potential parents who wish to have a healthy child may wish to have the foetus tested for a defective gene. At the present time these tests are performed on the foetus within the uterus by extracting a small amount of material for genetic analysis (Cadrin and Golbus, 1993). If the child is found to be affected the parents are left with the stark choice of aborting the foetus or having a child who may suffer from a serious and usually life-threatening disease. Potentially, developments in *in vitro* fertilisation technology may offer new options. One such concept is that eggs that have been fertilised in the 'test tube' could be genetically tested before being implanted into the mother (Wells and Sherlock, 1998; Lissens and Sermon, 1997). This testing would take place at a very early stage before the nervous system has begun to develop. Some studies in this area have already been carried out

(Wells and Sherlock, 1998; Lissens and Sermon, 1997). Potentially, such testing could be used to select embryos with other genetic traits such as eye colour. Although this may become feasible, it is likely the cost and difficulty of the procedures involved would stop it becoming widely used.

As the cost of Microarray technology falls it will become possible to screen an individual to determine if they carry any defective, life-threatening or even socially 'undesirable' genes. Although there are clearly medical uses for this technology it is highly likely that other parties would be interested in this information. Insurance companies are likely to be one such group. A knowledge of the genetic strengths and weakness of an individual would be of great use to companies who wish to set premiums advantageous to themselves. It is quite possible that some individuals could even be classed as uninsurable based on their genetics alone. Clearly this is an area where sensible legislation would be highly desirable to prevent genetic discrimination.

Gene therapy

The treatment of diseases with gene therapy has been the subject of numerous studies (Gottschalk and Chan, 1998). Potentially, this approach could be used for the treatment of inherited genetic disorders, such as cystic fibrosis, as well as being a potentially useful treatment for more common conditions, such as coronary heart disease, cancer and, perhaps, AIDS (Morgan and Anderson, 1993). Many technical difficulties still have to be overcome before gene therapy becomes a viable treatment option. In particular, effective ways to deliver genes to target cells have to be perfected (Takakura, 1996). This is partly because even a small piece of DNA is many times larger than a conventional drug. To date virus-based systems have been most extensively used to deliver genes to cells; however, concerns have been voiced over the safety of inserting viral DNA into humans. Non-virus-based mechanisms of delivering DNA to humans are being investigated and may be perfected in the near future.

One of the major issues produced by gene therapy is related to the type of cells that need to be treated to 'cure' a patient of an inherited genetic disease (Morgan and Anderson, 1993). At the most basic level there are two classes of cells that can be treated with gene therapy: somatic cells that make up the organs of the body and germ cells that comprise the male sperm and female egg. If somatic cells are modified then only the treated individual is affected. In contrast, altering the germ cells of an individual means that his or her offspring and descendants would inherit the genetic alterations that have been made. To date, therefore, gene therapy has been confined to somatic cell therapy, although arguments can be made to consider altering germ cells. Treating an individual early in life, perhaps as an embryo, would mean that an individual and their descendants would be effectively cured of the disease. It is likely that the technology for this approach will be developed; however, the cost of this technology is likely to limit its use for the foreseeable future.

Environmental release of genetically modified micro-organisms

The debate on the potential benefits, and risks, from the release of genetically modified micro-organisms into the environment is likely to continue. Numerous possible applications of genetically modified micro-organisms can be envisaged (Franco and McClure, 1998; Grasius *et al.*, 1997; Isken and deBont, 1998; Jordan, McGinness and Phillips, 1996; Sorokin, 1997). For instance a microbe that could degrade oil to biologically harmless molecules would potentially be invaluable at the scene of an oil spill. The use of an organism in this way would require its release into the wider environment. Once this had taken place, control of the genetically modified organism would effectively be lost. Therefore, before such a release, it is very important that the potential risks are clearly understood. One problem is that science still has to develop the means to model the global environment accurately. Numerous examples have demonstrated that the release of non-indigenous life forms can have devastating and totally unexpected results on the local ecosystem. To some extent this problem is amplified because genetically modified organisms are novel life forms that have not previously existed in nature. One can develop some idea of the potential risks by extrapolating the potential of a genetically modified organism to the worst-case scenario. In the case of a bacteria that could digest oil, it might be able to survive, proliferate and spread widely. All oil sources would potentially be at risk of destruction by such an organism. Although this may seem far fetched, it illustrates the fact that predicting how a novel life form may interact with the wider environment is extremely difficult. In general, it is reasonable to assume that if a life form is given a novel survival mechanism, such as being able to use oil as an energy source, it may establish itself in the environment. How widely this organism is able to spread and how it may interact with the indigenous life is unknown and presents a risk that must be considered before genetically modified organisms are released. It is likely to be some time before science can reliably predict how a genetically modified organism may interact with the environment. Until then the release of genetically modified life forms into the environment, where they cannot be controlled, is probably unwise and it is probable that they will be predominantly used in contained environments.

'Designer' tissues for transplantation

Use of cells and tissues grown in the laboratory for the treatment of humans is a likely growth area. Skin grown *in vitro* (in the laboratory) is already becoming a widely used therapy for the treatment of burns (Eaglstein and

Falanga, 1997). This approach is likely to expand as scientists become more proficient in growing tissues in the laboratory. Potentially, organs, such as hearts, kidneys and lungs, may be grown for transplantation. Whether this technology will use tissues grown exclusively in the laboratory, or those of genetically modified animals, such as pigs, will depend on numerous factors including the precise application required, the level of technical difficulty and costs.

Assessing risk from biotechnology

The need to assess potential risks to human health and the environment that might arise from the use of biotechnology will be of increasing importance. At the most basic level there are two methods of assessing risk. First a new situation can be compared to a similar event where the outcome in terms of risk is known. Alternatively, an experiment that models the real world can be performed. For instance one can pose the question, 'what are the likely results of a car crash to the occupants of a car?'. This question could be investigated by examining previous crashes where the outcome was known. In this case a pre-existing knowledge base is used to extrapolate conclusions about what may happen in the future. Alternatively, crashes could be recreated under controlled conditions. This would be an experimental approach. Experimentation is most important when the knowledge base is small or where a new situation is met. Even in this simple case there are numerous factors that can influence the outcome of a specific crash (for example car type, speed, safety features fitted to the car, the skill of the individual driver, etc.). These would probably need to be addressed either by examining a large knowledge base or by experimentation. When dealing with biological systems in the global environment the number of variables are far greater.

In terms of biotechnology there are potential harmful effects to both humans and the environment, and the knowledge base of some aspects of these effects is small. There is a large literature on toxicology and medicine that can guide decisions about the potential effects of genetically modified organisms on human health. There is currently far less information available about the potential interactions of genetically modified organisms with the environment, and more experimental evidence is required. Even when experimental data are available scientific opinion can be divided. Interpretation of scientific information is not always clear cut and different scientists may interpret the same facts differently, depending on individual background, training and experience. Differing scientific opinion is usually most apparent early in the scientific process where there is a small amount of data that can be conflicting. In particular, isolated single studies should be treated with caution. It is not until several studies show similar results that most scientists would feel comfortable in accepting a finding. Over time, as studies accumulate, a consensus is reached in the scientific community where the majority of workers agree on a common interpretation of the available data. However, even at this point, surprising findings can arise which may result in a complete change in scientific opinion.

How this science develops should depend as much on public debate and sensible legislation as on funding priorities and the interests of researchers and industry. It is increasingly becoming vital that scientists learn to communicate effectively with the general public, who in turn, need to be informed as to the issues. Without this two-way communication, sensible, informed and beneficial debate about the progress of biotechnology will be hampered.

Notes

The author would like to acknowledge Marie J. Lucey, B.Sc. for help with computer graphics and Sari L. Goldberg, M.Ed. and Helen L. Ross, Ph.D. for critical reading of this manuscript.

3

Biotechnology and international relations: the normative dimensions

Hugh Dyer

Introduction

Consideration of the normative dimensions of biotechnology in the context of international relations rather implies a point of confluence at which these two otherwise distinct areas engage with one another. While there is, perhaps, some such location in the institutional structures of governance and exchange, it appears that engagement is less than constructive – more a case of mutual aggravation. There is common ground, but neither is this a happy place – the steamrollered terrain of growth-oriented economic globalisation. Here, the normative issues of international relations, which centre on distributive justice in relation to political and economic rights, engage with distributive justice in relation to the benefits and burdens of biotechnology, in which issues of political control and economic access are central. At root there is a common denominator in concern about human development only recently (in historical terms) overlaid with largely instrumental concern about the natural environment. Hence this chapter must begin by addressing the tension between traditional anthropocentric or homocentric (human-centred) ethics and ecocentric or biocentric (ecosystem-centred) ethics (Eckersley, 1992; Fox, 1990). Rather than seeing only the economic potential of biotechnological manipulation of nature for the purpose of human development, a 'green' theory of value, for example, would suggest (among other things) that nature is valued for its naturalness, and so any kind of intervention is devaluing. It thus stands in stark contrast to socialist or capitalist development assumptions (Goodin, 1992). So, while at the heart of the normative issues in biotechnology is the type and degree of intervention in the natural order (the human life cycle being an especially sensitive case in point), the political environment, in which such values are expressed and debated, is at once united in its anthropocentrism and fragmented by particular interests. In this latter respect most obviously so in international relations.

Values in motion

Because environmental values have become increasingly important in international relations, it is necessary to explore the significance of such ethical positions with respect to international obligations and co-operation concerning biotechnology. More importantly, if the international system is changing or is likely to change with the impact of biotechnology, it is imperative to consider the ethical parameters of any new collective understanding. 'Collective' is intended here to suggest some global cosmopolitan compass. Yet it is not to be expected that such ethical parameters will be directly constraining, or even that they will be widely acknowledged in particular local instances. However, some such ethical framework will inevitably emerge as political values are exchanged (that is, contributed to or borrowed from a global 'pool' of values) through processes that emphasise the interconnectedness of local problems and the global condition. The development and distribution of biotechnology is an example of such a process. This view seems to be consistent with the broad and fluid notion of globalisation employed by Robertson (1992), emphasising trends of cultural 'globality' and 'unicity' in which global culture remains politically contested even as globalisation is increasingly structured. Values informing the deployment of biotechnology, as with other political values, will influence such local and global political cultures, which in turn inform the politics of identity and difference – that is to say, world politics (Connolly, 1989).

Values invoked by the advent of biotechnology contrast with those of the states system and thus present new challenges and possibilities for international relations. This dichotomy of values underlies a contradiction between traditional accounts of international relations and the emerging agenda, since biotechnology is not readily constrained by the boundaries of the territorial state. Furthermore, 'arguments about the merit or appropriateness of biotechnology are made within a value system different from that used to judge other technologies' (Gendel *et al.*, 1990: 343).

In presenting anthropocentrism in biotechnological ethics as a norm, the argument suggests that it is a social product which entrenches a particular value perspective, or stand point, in relation to biotechnological issues. The society in this case is the broadest possible one, though it matters whether this is understood as international society (a society of states) or cosmopolitan society (global civil society). It may be argued, for instance, that the influence of civil society and non-state actors in the politics of biotechnology parallels other processes which transcend the state system, such as the development of international financial markets. A normative approach to these processes allows a more subtle interpretation of international politics, including the common denominator of environmental concern and the potential risks of biotechnology.

Ethical anthropocentrism or homocentrism, in accounts of human relations to the environment, is contrasted here with the alternative ecocentric or biocentric approach. The principal thrust of this distinction is to illuminate the modernist conception of the non-human environment as 'nature', which underwrites resource exploitation in the interest of *human* progress, and the manipulation of nature through biotechnologies (Oelschlaeger, 1991). Through the device of mind–body dualism, this externalisation of 'nature' can even be extended to the corporeal existence and function of the sentient moral agent who holds such a conception. This conception is reflected in both ethical and political systems to the extent that they, respectively, encourage the individual pursuit of material wealth, and distribute (on some basis) the proceeds of apparently unlimited economic growth including that in the area of biotechnology. The alternatives suggested by ecocentrism are not necessarily antithetical to such human aspirations. The human species is part of the total ecosystem, or biosphere, but human impact on the global ecosystem is a central concern, and the implications of any shift towards ecocentrism are profound. More specifically, the connection between appropriate and acceptable ethical values and similarly acceptable political values is germane to the problem of the separation of ethics and politics. This issue is brought out in the case of biotechnology because of the tension between the existing and accepted political conditions in Western societies (for example, individualism, property rights, separation of state and society, market economies, the territorial state, etc.) and the ethical challenge posed by biotechnological innovations, such as genetic engineering. While existing political structures can cope with the requisite bargaining and distribution of burdens to a degree, there are limits to the abilities of existing political systems which cannot be extended without reconsideration of the underlying ethical life of the participants. Even in non-Western societies where property rights have been conceived differently, for example in terms of inter-generational group ownership of land with conditional possession rights for individual users (Omari, 1990), the introduction of capitalism and other Western practices has undermined traditional ethics and created similarly limited political systems. It may be that an appropriate biotechnological ethic is to be found outside of the Western tradition (Khalid, 1992). This in itself would introduce new lessons into the predominantly Western discourse about, and Western historical origins of, the existing international system as well as informing ethical debates about the role of, for example, sources of traditional medicinal remedies and their modern technological derivatives or surrogates.

Furthermore, the problem is compounded when political systems are understood to be without ethical content – ethically neutral. This is not to say that ethical choices are not being made, by individuals and communities, but simply that political rights and duties are thought to fall into another category. Importantly, political rights and duties have traditionally arisen in the context of the state, whereas issues like biotechnology imply both a limitation on the ability of individual states to defend rights or meet duties (states alone can have little influence on the control of biotechnologies) and at the same time a potentially broader basis for normative ethics (morality) in a global society. Where politics and ethics are distinguished institutionally (in theoretical traditions or in political practice) there are structural barriers to reconstructing politics in the light of ethical life. These structural barriers can be overcome by breaking down the distinction between ethics and politics – acknowledging the ethical aspect of politics and the political aspect of ethics – and transcending the strictly international to consider also the global-local dimension, ethically and politically (Beitz, 1983). It is important in this respect to adopt a broad understanding of ethics, not as a commitment to natural law or religious belief, since these cannot easily be re-forged into theory (Kolakowski, 1988), but as an acknowledgement of the normative dimensions of political interaction. Equally, it is necessary to aquire a full understanding of the increasingly global context of political interaction, not conceived as simply 'world-wide', but rather as the relationship between local issues addressed in a global context and global issues addressed in a local context – local biotechnology issues (say, involving proximal land resources) may demand a global strategy, and global problems (the control of biotechnological products) may demand local action.

International politics in play

The international dimensions of biotechnology concern the impact of these issues on traditional conceptions, theories and practices of international politics. Naturally there are two sides to the story suggesting, respectively, that all is generally well in international relations and existing systems and institutions can cope, or, that the existing system is essentially flawed and in need of radical overhaul (or at least significant adjustment). In the area of international law, it has been argued, traditional sources of law, such as treaties, custom and general principles (Brownlie, 1990), are well tested and can continue to serve in dealing with the broad area of environmental problems. However, the doctrine of sovereignty may have to be conditioned in some respect to ensure environmental responsibility (Birnie, 1992). The many and various formal institutions of international politics (for example, the UN) may be well suited to dealing with emerging environmental issues, but even they have had to adapt (UNEP, UNCED, the Convention on Biological Diversity, etc). Economic institutions, such as the World Bank, have been criticised for not adapting quickly enough, and are in the process of reform to ensure that the projects supported are more than just economically viable but environmentally responsible as well (Rich, 1994). Others argue that existing economic structures are sufficiently powerful and adaptable that future environmental degradation and biotechnological risks can be mitigated by modest refinement of market forces. Thus it is appropriate to discount future income (under conditions of economic growth) thereby discouraging excessive investment in environmental returns (Beckerman, 1992). Henry Shue (1992) points to the injustice of burdening developing states; having at best played a small part in the industrialisation that is largely responsible for environmental degradation and the new biotechnologies, they require their existing resources (and more) to combat the future effects of change.

Novel global issues, such as biotechnology, present special problems for theory and method due to uncertainties and non-linear relations of cause and effect. The principal questions, for social science, include what people do in relation to biotechnology (Sayler, Sanseverino and Davis, 1997), how people are affected by biotechnology (Shiva and Moser, 1995) and what information people use in making choices about biotechnology (Young, Sterna and Druckman, 1992). At each step, socio-economic and scientific value structures exert their influence on perceptions of both the problem and its solution. Social value structures are readily understood in terms of ethics, but, more importantly, the familiar features of ethical relations can also be seen in knowledge structures (science) through the privileging and protection of certain standards and assumptions in our epistemologies, as much as in our societies (Breyman, 1993).

At some point such scientific and social assumptions must overlap, and it is here that the values informing the international relations of biotechnology are likely to reside. Where there are considerable differences in social attitudes, engaging in the politics of biotechnology is bound to be difficult even if there is epistemological consensus, such as in the ecological 'epistemic community' influencing environmental politics (Haas, 1992). We should note, for example, that there is disagreement about perception, estimation and management of risk, including concern about the environment, suggesting that the problems of risk hinge on knowledge about the future and consent about the most desired prospects (Douglas, 1982). Where these are uncertain and contested, there is no obvious solution, and, interestingly, 'any form of society produces its own selected view of the natural environment' and 'common values lead to common fears' (Douglas and Wildavsky, 1982: 8). One group of authors addressing the question of acceptability of risks notes that

> That choice depends on the alternatives, values, and beliefs that are considered. Values and uncertainties are an integral part of every acceptable-risk problem. As a result, there are no value-free processes for choosing between risky alternatives. The search for an 'objective method' is doomed to failure and may blind researchers to the value-laden assumptions they are making. (Fischhoff, Lichenstein and Slovic, 1980: ii)

This seems clear enough, when applied to national societies, which will not be equally or similarly affected by biotechnologies – for example, some will find it easier to intervene or adapt than others, some will feel greater economic pressure, the demand for more hardy, disease resistant and vigorous crops or more productive dairy herds will vary, and competing environmental issues may take precedence. Small island states, for example, will be most concerned about rising sea-levels, and dry equatorial states will be concerned about desertification, while others may be concerned with water and air quality, or rainforests. However at the international, or more appropriately the *global*, level world society will have to contend with the social biases of its constituent groups while contemplating the global dimension of these problems. There are clearly some such problems (for example relating to regulatory regimes, markets and finance, etc.) which cannot be properly understood at any other level. In this transition from the local to the global, perceptions will continue to be informed at the most fundamental level by ethical relationships, even as these relationships extend beyond local or national social boundaries and become transboundary or global issues. Thus are the familiar questions of international ethics (just war, positive peace, justice in procedural and distributive arrangements) touched by the more recent complication of biotechnology, which requires us to consider not only the relationships between nation-states (if seen as distinct and separate ethical spheres) and between people but also our relations with the non-human world as well – with the total environment. Beyond the tensions between change and continuity, centralisation and decentralisation, which Rosenau discusses in terms of 'turbulence' (Rosenau, 1993), world politics may also be characterised by a tension between the anthropocentric concern with political and economic community and the ecocentric concern with an integral ecological ethic. This latter could transcend present conciousness (and perhaps alter ecologically damaging political practices). It suggests a kind of 'post-anthropocentric cosmopolitanism', which is ecologically inclusive rather than exclusive (Kealy, 1990: 87–100).

Biotechnology in question

Hence we can see that the definition of biotechnology is open to bias, as noted by Manning (chapter 2) on the science of biotechnology. Nevertheless, the issues that are of pressing concern at the end of the twentieth century arise from the specificity, and genuine novelty, of such technologies as genetic engineering (as opposed to mere 'breeding') through which the possibilities of genetic manipulation have been qualitatively enhanced. Much of this technology is already embedded in our political economy (monoclonal antibodies have been big business since 1975). The source of anxiety, and of ethical dilemmas, is the extent of uncertainty about future technological developments and about the consequences of deploying even current technology. Introducing genetically modified organisms into a 'new' environment is inherently unpredictable, due to the potential for disruption of complex ecosystems by living, reproducing (possibly aggressive) entities with a survival imperative. Such certainty as might be sought from scientific sources is diminished by a reliance on very specific scientific expertise in biotechnologies, when the context of ethical and political concern is an holistic ecology (the planet, not the laboratory). This is further exacerbated by contestation about the science within the scientific community itself. In this atmosphere, the usual problem of entrusting or delegating ethical and political decisions to scientific experts is made more obvious. We can see in this a parallel with the 'Atomic Age', in which the competitive arms race and space race were the special province of technical experts and political control was dependent on specialist knowledge. In the case of biotechnology there are more actors (not just the military–industrial complexes of states) and less control, even if such mechanisms as the Cartagena ('Biosafety') Protocol of the Convention on Biological Diversity (see Vogler and McGraw, chapter 8, this volume) provide a further parallel with the anxiety of an earlier period about how to control a recalcitrant technology. This case is also marked by the additional control problem of easy replication and dissemination of technological information (for example, genetic codes) because of the parallel development of the communications revolution (particularly the internet). As always, science tells us what is technically possible, not what is politically and ethically desirable, nor how political and ethical choices might be given effect once the genie is out of the bottle. In these circumstances, it is as important to know something of scientific values as it is of political or ethical ones. Without reducing the discussion to epistemology *per se*, the close relationship between epistemology and ethics should be enough to alert us to the importance of the politics of knowledge.

From a gender perspective, Bretherton and Stevenson (chapter 4) raise just such concerns. Here we see the consequence of a separation of those 'in the know' from the general public (technocracy versus democracy), and the objectification of both women and nature by science. This is knowledge as authority (Breyman, 1993). In spite of the general trend towards masculinist control of nature, and the universality of Western science, this also points to the cultural specificity of knowledge – a gender culture is this case, but the way is open as specificity arises from the source of authority. So it is that an ethical context, a set of values, necessarily surrounds and impermeates what might otherwise be a purely scientific debate on biotechnology. The central issues in the life cycle are thus not just given instances of birth and death (which carry with them, in every case, ethical dilemmas every bit as troublesome – definitions of life, and death; quality versus quantity of life; etc). There are also the more encompassing issues of the commodification of life, the manipulation of life (eugenics), the social dimension of reproduction and fertility, the provision and consumption of food, all of which can now be touched by biotechnological innovations. Consider the example of assisted reproduction (an expensive 'technological fix') in a world of too many suffering children, or that of GM crops in a world where indigenous traditional farming is already greatly threatened. The technology itself could be potentially value neutral were it possible to escape the political and ethical context in which there is discrimination among the recipients (or victims) on the basis of race, sex, religion, age, wealth, knowledge, etc. But such an escape is not possible, and science can not be value free. There may, of course, be alternative values informing science: 'biotechnological optimism', as against 'nature enhancement', or simply the protection of nature from degradation or of women from reproductive abuses. In all these contexts there is scope for political intervention or sponsorship (not least by states), and here the more recognisable concerns of international relations come into focus as nations, peoples, and identity are at the root of international politics (and ethics). Whether we are now speaking of technology or of politics is precisely the question; if technology is defined in terms of human use and intent, then we can not easily separate the two – nor divorce them from ethics.

Normative dimensions at work

Thus definitions of, and assumptions about, the risks of biotechnology come to bear on ethical considerations. For example, the threat to biodiversity raises two sets of ethical issues (anthropocentric and ecocentric), the first concerning human interests and the second concerning interests of other species and of ecosystems as a whole. As exploiters and consumers of the non-human environment (for food, fuel and other industrial commodities), humans have an interest in environmental management which brings with it ethical demands in respect of other humans, including future human generations which will benefit from biodiversity (anthropocentrism). Yet these consumptive activities also require that humans consider their ethical relations to other species, and are aware of the role of the human species in the planetary ecosystem (ecocentrism). This latter concern is most commonly observed in respect of familiar creatures (pets, farm animals, the victims of blood sports, etc.) and designated terrain (farms, gardens, parks, forest reserves, etc.), but must logically extend to all other species and the ecosphere as a totality, since there is a high degree of interdependence and an element of balance elegantly managed by 'nature' (which we may recognise as being crudely emulated by humans in 'state of nature' approaches to international relations).[1] This extension of concern should follow even from an anthropocentric view, since humans are beneficiaries, and yet the objects of concern will not be fully integrated into human ethical parameters without an ecocentric acknowledgement of all species and ecosystems as intrinsically valued in their own right. This is well illustrated by contrasting approaches to controlling agricultural pests, represented by biotechnological 'fixes', on the one hand, and ecologically sound innovative practices on the other.

Even so, it is inevitable that an anthropocentric approach to ethics will have to consider the knock-on effects of exploitation and consumption just because the human environment is part of the total environment. Apprehending the natural environment as a distinct entity, objectified in relation to the human subject, institutionalised as property by human societies, is simply to miss this obvious point about the conditions of life in a biosphere. These conditions are both the cause and effect of the globalisation of human practices, which forces the issue of 'self versus others' / 'insiders and outsiders' / 'us and them', since such distinctions are qualified by global environmental changes which do not allow protected (protectionist) domains or exclusions.

Consideration of biotechnology also exacerbates some of the difficulties encountered in a more conventional account of international ethics. The various issues in North–South relations which have for so long occupied (and occasionally dominated) the international political agenda are now overlaid with biotechnology issues, and the mutual attentiveness this brings is an aspect of globalisation. Biotechnology and other global environmental problems, in simple terms having their origin in the industrialised north and potential resolution in the developing south, add new stakes to the bargaining over distribution of resources and opportunities (Redclift, 1998). The debate over sustainable development is of general concern, having implications for both developed and developing states. The long-standing issue of population pressure, often measured in terms of the planet's capacity to sustain life but obviously unequal in its distribution, raises the thorny issue of legislating human life – should governments anywhere or some international authority be allowed to intervene, and if so, how? Furthermore, the intervention required may not be the most obvious one (biotechnologies for population control, or 'family planning'), since there is evidence that population pressure arises from socio-economic practices and previous environmental degradations which encourage population growth as a means of survival (Thompson, Ellis and Wildavsky, 1990). Something may be gained from maintaining a degree of balance in the ecosystem, so that populations can find their own viable stable state in harmony with their environment, without the necessity of biotechnological interventions.

In the international political economy, technology is a driving force. Biotechnology in particular has created the most radical and tempting possibilities, naturally seen in this light as commercial or economic opportunities. Here technology is both in and of economic and knowledge networks – as Russell (chapter 6) says in his chapter, the networks are composed of 'technological actants'. The issue for this chapter is whether these 'actors' are also 'moral agents'. In the discipline of international relations 'actors' is an essentially neutral term referring to any political player (state, NGO, MNC, etc) which fits into the structure of political power. It is a useful analytical tool for avoiding value questions, and yet where these can not be avoided the moral as well as political agency of any actor must be considered. For most economic issues, the ethical dimension emerges in the form of distributive justice (or injustice, as the current distribution of biotechnological knowledge, ownership and control suggests), but if maldistribution is attributed to the economic system itself, the moral responsibility is diffused by attribution to an amoral and unaccountable structure (seen as a cause) rather than to accountable decision-makers (who usually find immunity within corporate identities, including states). In this way the structure–agency debate helps to inform our understanding of biotechnology as either an unchallenged product of the technocratic structures of the international political economy or as an ethically sensitive area which demands that efficacious decisions be made by morally responsible agents, and that they be held accountable for those decisions. If the latter view is taken, we should expect biotechnology to be a constant item on the political agenda, since the non-linear character of technological development means that specific decisions may have limited application even if general points of principle may have a longer shelf life – the precautionary principle stands out here since uncertainty, both scientific and ethical, is the dominant characteristic.

International relations in controversy

The implication for the study and practice of international relations is that most of the theories and methods that have been devised to cope with a system of territorial states will have to be modified in some respects to cope with the global character of the biotechnological revolution. Particularly, local traditions and practices will have to find some location within the broader global context through the politics of identity and difference, which amounts to the globalisation of politics as distinct from exclusivist international politics. The increasing importance of biotechnological ethics is potentially both a cause and effect of such changes in global political culture, which is not to say that 'locality' is to be subsumed by 'globality' but rather that each provides a context for the other, as globality emerges as a late-modern phenomenon, or indeed as biotechnology itself drives us into a post-modern condition (Wheale, 1994). It appears that while fundamental value changes are inherent in processes of modernisation (Inglehart, 1997), and these may support awareness and concern for environmental issues such as the introduction of biotechnology, there is still room for controversy.

With Vogler and McGrew (chapter 8) on the international regime for biotechnology we see that, while 'biosafety' is key to the Convention on Biological Diversity, the precautionary principle is downgraded to the precautionary 'approach' through the usual process of watering down international agreements in the hope of reaching consensus or unanimity.

Furthermore, while biosafety is an important and controversial issue (still unagreed, as the bracketed text shows), the general tenor of the overall convention leans towards concern with trade and intellectual property. In the regime context, Vogler and McGrew define norms as behavioural injunctions, and the precautionary principle qualifies as a means of justifying prohibitions where prior informed consent is not possible given the absence of certain knowledge or the presence of uncertainty. The normative implications in this sense hinge on two issues. One is the notion of informed consent (as for medical procedures), where the ethical issue is whether a genuine choice is being made by the relevant moral agent. The other issue is prudence, or the avoidance of unnecessary risk, as implied by the imperative of taking precautions. Here we are directed to the harm that might be caused by culpable negligence, rather than by a wilful act of violence or deception. All of these ethical pitfalls are exacerbated in international relations by the demands of sovereign autonomy and economic growth (expressed through their joint product – the competition state) which run counter to the precautionary principle, and indeed raise ethical issues of a different kind, such as self-determination, autonomy, 'the right to develop', the necessity of employing and feeding and providing medical care to populations, etc. As with the norms of all international regimes, and normative ethics (morality), there will always be a contrary set of values giving rise to contrary political demands, even in the same space and time. Certainly across space and time we can see differences. Falkner (chapter 9) notes the cultural variations between Europe and the United States concerning GM foods. The 'greyness' of these issues and the mix of costs and benefits is not perhaps unique, and we can point to long-standing tensions between the developed and developing world which arise at virtually every step of technological and economic change this century. What remains unique about biotechnology is the immediacy of its implications for human life and the novelty of its science.

Finally, therefore, it is necessary to consider to what extent existing approaches to international relations need modification. As we have seen, in some respects biotechnology problems merely compound existing difficulties, while in other respects these problems have forced the question of whether our existing ethical and political relations are up to the challenge. In the former case we may be content to apply well-known consequentialist arguments arising from prudential agreement based on mutual interests between nation-states. But the argument from rational self-interest begs the question of how any national interest can be distinguished from the 'interests' of groups of people, and of the biosphere (assuming human values derive at least in part from experience of life as inhabitants of the planet). This is the usual objection arising from an holistic approach to the self and the community, or the particular and the general, now extended also to the local biotic community and the global. Similarly, we may wish to apply deontological (categorical) arguments concerning the intrinsic value of nature and hence prohibitions on biotechnological intervention. For anthropocentrists, this line of argument is conditioned by the view that the intrinsic value of nature lies in its ability to serve the interests of humankind (including national interests, usually defined in economic terms). For better or worse, normative theorising must ground itself in social values, whether laudable or deplorable (though we may still wish to encourage the former). However, ecocentrists argue that we may cultivate an improved human vision and reformed social values: after all, the foundations of political theory and practice are not static.

The pragmatic solution to this quandry is simply to mitigate against the worst effects of biotechnology and adapt practically and intellectually as necessary, and yet there is a degree of urgency derived from the finiteness of the planet and the irreversibility of some changes. The 'precautionary principle' is often employed to capture the essence of the dilemma – that since we do not yet know with certainty what the consequences of acting or not acting are, it is only responsible to act appropriately anyway. This is, responding to the worst-case scenario (for example, bans on the introduction of GMOs) – a familiar strategy in the area of security studies – but of course a traditional role for ethics (and I would claim it also for normative theory – not conflating this with ethics, *per se*) is to address the prospects for the *best*-case scenario, that is to say the conditions of the 'good life'. Human aspirations cannot be ignored, but we may also discover that the aspirations which have led us down a dangerous path of materialist economic growth are only manifestations of prevailing conceptions of the good life and are subject to change. These factors alone demand some degree of modification to conventional thought and practice, and indeed such changes are already visible in many societies and in the agenda of international politics.

The security implications of biotechnology are perhaps the most dramatic, or at least most frightening. Biological weapons are not very new, but, as before, the novelty of modern biotechnology has brought qualitative change. The ethical issues are not new either, as we are now not speaking of a potentially benign technology gone awry (though Spear's point about dual-use technologies suggests an element of this). Here the ethical weight resides in the scale of destruction, rather than in the simple act of it, which invokes a horrible ethical calculus based on casualties. The ethical problem is similar to that of nuclear weapons, and of nuclear deterrence as a policy: the existence of the weapons offends (since they ostensibly have but one purpose) and yet the absence of deterrence may expose millions to oppression or destruction. In a further similarity, the horrible nature of the weapons encourages a control regime, which can be interpreted as a limitation on distributive justice by denying equal access to this biotechnology, and doing so on the basis of a necessarily subjective ascription of political responsibility or stability – both ethically suspect positions, but to be weighed against the perceived risk to human life. As Spear indicates (chapter 13), biological weapons are a levelling technology in military terms – a 'cheap and cheerful' alternative to fighter aircraft and nuclear submarines, and even more intimidating if in the hands of those perceived as irresponsible. The wide dissemination of this technology because of dual uses (civilian and military – again, as with nuclear products) leads to ready access, and this is exacerbated by relative ease of 'weaponisation'. While more such capacity could hardly be viewed as a good thing, it remains that there are those who may feel the necessity of resort to such weapons, and may even be able to find some ethical position to support that choice (self defence, for example). As Dando (chapter 10) and Littlewood (chapter 12) note, the science is outpacing political control, and with highly specific biotechnologies allowing ethnic targeting this is surely an urgent matter (with Kosovo fresh in the mind, the prospect of biological weapons in that context seems all too real). Yet even this is not ethically uncomplicated, since as they also note in relation to the future of biotechnology, the mechanisms of control must stand up against the demands for free trade and confidentiality of proprietary information. Here the pursuit of security seems justified, and yet we would wish to distinguish between commercial security (protection of property), economic security (of markets), national security (of states), and international security (potentially of humankind, though not perhaps of nature as a whole) – as always when security is invoked we should pause to consider who the beneficiaries of security are, and what the ethical implications of one group's security may be where others are concerned.

Conclusion

Successful responses to biotechnology may require rather more effort than we are presently willing to contemplate, including perhaps even the adoption of a distinctively biotechnological ethic deriving directly from relations between humans and the global natural environment, rather than an ethic derived from the somewhat disreputable arena of relations between nation-states (Warner, 1992). Political practice will no doubt continue to respond reactively to the many and various demands arising from biotechnology, and it is to be expected that traditional methods and tactics will be employed in the pursuit of state interests, and corporate interests, as much as is possible. However, we need not look very far for evidence of three complementary trends: the first is a general trend towards globalisation, which is largely independent of state behaviour, whether in financial markets, trading patterns, communications or culture; the second is increasing political and economic integration and cooperation; the third is the explicit impact of biotechnology and other environmental issues on institutional policy and practice, including international organisations, all levels of government, non-governmental organisations and industrial and other private sector organisations. In all cases, individuals are also implicated as the motivators or the motivated, and in this context: 'The issues of how we treat one another and the world around us in a responsible and ethical way cannot be divined through nature. They are human issues and depend on human values. The choices have always been with us, and were not created by biotechnology' (Grace, 1997).

These circumstances suggest that the international relations of biotechnology are, and will continue to be, conditioned by emerging values and value structures, which are largely unfamiliar to the traditions of international relations and perhaps also to domestic political traditions, which defer to technocrats on technological matters, and possibly in some extreme cases antithetical to both. Yet ultimately the issues of biotechnology remain within our human grasp providing there is sufficient political will to reach out and engage with them – and, on balance, there probably is.

Note

The Hobbesian state of nature pits all against all, in an analogy of the anarchical realm of politics without central authority, and so gives rise to theories of international relations which suggest that a balance of power provides stability in anarchy. Of course, Thomas Hobbes, in his book 'Leviathan', was offering a *post hoc* justification of central authority (sovereignty) in the nation state, while international relations concerns itself in the first instance with relations between established sovereign states. It transpires that relations between sovereign states are something less than anarchical, given the complexities of interdependence, that states are something less than sovereign and that the balance of power calculus is something less than 'natural'.

4

Ambivalence and anxiety: women and the techno-scientific revolution

Charlotte Bretherton and Karen Stevenson

> molecular biology assumes that life, at base, is the end result of the properties of the materials from which it is made, and that we can understand it by treating it as a collection of complicated molecules acting together rather like miniature clockwork. Life, it says, is a molecular machine. (Bains, 1987: 6)

> We are on the brink of a new revolution of quite awesome power. The revolution in molecular biology will give us the ability to divert and control human evolution to an unprecedented extent. It will enable us to manufacture new life forms to order, life forms of every sort. (Harris, 1992: 5)

These statements – written, respectively, by a (male) biochemist and a (male) philosopher – are illustrative of a reductionist, control-oriented approach to science and its application that has long been the subject of controversy among feminist scholars. Positions in the debate have ranged widely, from rejection of post-Enlightenment scientific thought as inherently life threatening (Merchant, 1982) to a focus upon women's access to new technologies as an issue of distributive justice on a global scale (Mitter, 1994). However, it is the arrogant assertion that we will 'manufacture' (with its connotations of factory farming) 'life forms to order' that has caused such recent trepidation among theologians, the media and the general public alike. In suggesting the potential for finally unravelling the secrets of life and death, the claims of biotechnology are uniquely challenging. They appear to be, simultaneously, both life enhancing and life threatening – in that they juxtapose a yearning to produce beautiful, healthy babies, or to achieve eternal youth, with deeply rooted fears that the life-giving roles of God, nature and women themselves are being usurped by male techno-science.

While the gender implications of modern biotechnology are wide ranging, this chapter focuses primarily upon an area which is not addressed elsewhere in this volume – that is, human biotechnology and the attempt of masculinist science to control the very processes of life and death. In a global system in which there continue to be substantial inequalities of gender, race, wealth, information and longevity (among others), the 'benefits' of biotechnology are certain to be unevenly distributed. Indeed, some have feared the creation of 'a gigantic slaughterhouse, a molecular Auschwitz in which valuable enzymes, hormones and so on will be extracted instead of gold teeth' (Professor Chargaff, in The Genetics Forum, November, 1993).

In our focus on human biotechnology we consider first the denial of death in modernist culture, and the increasingly sophisticated biotechniques that hold out, at least for the wealthy, the promise of immortality. Man's (sic) attempt to overcome the limitations of nature highlights the ambivalence many women feel towards a (masculine) technology that considers itself to be superior to, and separable from, a (feminine) nature. We will argue that, in both matters of life and matters of death, two major implications of the new biotechnology have global significance. Firstly, the commodification of human life associated with the intrusion of market principles into matters of life and death, to the extent that the parts and processes of humanity have become saleable items on the open market. Organs, sperm, surrogates, eggs, all can now be purchased; indeed, there are internet sites devoted to their sale. Secondly, a related effect of commodification is the revival of eugenicism. In an open market economy it is already evident that the parts and processes of some humans are accorded more value, and are thus more desirable, than others. This, we suggest, can only be exacerbated in a future in which our genetic heritage becomes both more fluid and more significant.

Male techno-science and the subjugation of nature

Although science and technology still function as signifiers of 'progress' in modern industrialised societies, there has been a growing disillusionment with techno-scientific interventions in recent years. This loss of faith in the idea of unfettered progress is evidenced by the ever-growing public campaigns against untested and feared technologies. From the wide-ranging anti-nuclear campaigns of the 1960s, 1970s and 1980s to more recent interventions expressing fears over cloning, gene therapy, post-menopausal mothers and genetically modified foods, the public has indicated a desire for science to be held more accountable. Loss of faith in science, technology and traditional models of development is also indicated by the South's critique of economic imperialism, the green's critique of Western industrialisation, the feminist's critique of reproductive manipulation and the post-modern's critique of the basis of modern knowledge. What unites these separate critiques is a focus on the problematic nature of the hierarchies fundamental to Western dualism. The patriarchal capitalism of the industrialised world, it is argued, has been charged with the globalised subjugation of the natural sphere in its 'colonisation of women, of "foreign" people and their lands; and of nature, which it is gradually destroying' (Mies and Shiva, 1993: 2). The threat to nature can be seen as resulting from the hierarchical separation of culture (mind) from nature (body) as well as the hierarchies in Western patterns of knowledge (reason/emotion, expert/vernacular, objective/subjective). Thus, while in pre-humanist times 'nature' was a force to be reckoned with, appeased as well as appreciated, modern nature is perceived as pliant. Positive narratives of man's technological and scientific harnessing of nature and her resources imagine man as master of all he surveys and nature as the passive subject of discovery, a raw material to be manipulated and made useful through man's ingenuity. In such a discourse, women and 'natives' are aligned with nature – and are perceived to be in need of civilising by cultural and technological authority.

The taming of micro-organisms by biotechnology significantly extends this control of nature. Although biotechnology is not a new science – we have used micro-organisms such as yeast, for example, to change one thing (flour, grapes) into another (bread, wine) for centuries – major advances have been made in recent times. The discovery of the structure of DNA in 1953 and the knowledge that a cell's function could be identified, and potentially manipulated, made possible modern biotechnology's incursion into genetic modification. Optimists argue that biotechnology signals the end of disease, hunger and pollution, as the natural biological processes of microbes and of plant and animal cells are harnessed for the benefit of humanity without waiting for the slow and uncertain processes of crossbreeding or natural evolution. Nevertheless, the ability to eliminate or introduce specific characteristics in an organism, within a single generation, raises important ethical issues. And it is here that our ambivalence about the benefits of such incursions begins. We are now capable of manipulating life itself; of creating, or cloning, new plants and animals; of combining the DNA from different species; and of making decisions that, quite literally, influence life and death. Moreover, while we have gained science, it is claimed we have lost God. This has led to an 'existential homelessness' in Western thought, in which human identity, as fully alienated from nature, has 'no true home in it or allegiance to it' (Plumwood cited in Mellor, 1997: 113). 'Modernity, despite its pride in throwing off the illusions of the past, has not provided an earthian identity which gives a life affirming account of death, or comes to terms with death as part of the human condition' (*ibid.*).

Death and denial: the illusion of control

In the West there is a long-standing belief that there is a 'moment' of death, the moment at which life ceases and the soul or essence of the dead person is no longer.[1] Modern science not only assigns a time at which death occurred, but also a reason why it occurred at that moment. As Nuland (1994: 43) points out, we no longer allow people to die of old age, to just wear out: 'Everybody is required to die of a named entity, by order not only of the Department of Health and Human Services but also of the global fiat of the World Health Organisation.' The problem here is that, by assigning a *cause* to what may be a natural biological state, we reinvent an inevitable human process as a pathology to be cured. For if we can isolate the cause of death, at the same time we isolate the enemy that has the potential to steal human life and, implicitly, suggest ways in which such an enemy – and so nature itself – can be denied. This orientation of Western science thus illustrates the 'difference between a world view that recognises the inexorable tide of natural history and a world view that believes it is within the province of science to wrestle against those forces' (Nuland, 1997: 58). However, while Western science has succeeded in increasing life expectancy significantly over the last hundred years, morbidity figures in many regions of the world remain depressingly high. Thus, in much of Sub-Saharan Africa life expectancy is below fifty, little more than half that of countries such as Japan (79.9) and Canada (79.1) (United Nations Development Programme, 1998).

These global inequalities in health and morbidity are also evident within societies and increasingly reflect the availability of death-defying technologies for the affluent and the neglect of basic health care and nutrition for the poor.[2] The affluent, for example, have changed their lifestyle to the extent that heart disease is no longer the principal cause of death in the UK. So, while heart disease figures are still worrying, they also hold out the promise of control by enabling us to reflexively re-order our lives in the light of new medical knowledge.[3] Moreover, sophisticated bio-engineering techniques are increasingly promising add-on high-tech solutions to our physical shortcomings. Cloning technology now allows us to create new arteries using human rather than pigs' genetic material; electronic pacemakers safely trigger a predictable and steady heartbeat; mechanical defibrillator, directly implanted into the patient's body, reassert control in response to an irregular rhythm; narrow blood vessels can be widened or regrown; blood can be re-routed around obstructions with a bypass operation. In the most extreme cases the entire heart can be replaced with a donor organ or even a mechanical pump. Vat grown organs, again using cloning techniques, are just the latest borg technology.

The differences between the idea of heart disease as the body's terminal event and as the cause of death are not merely semantic – our faith in the promise of biotechnology, and our fantasies of control over natural processes, appear to escalate with each new step. Why not? Advances in science over the last hundred years have uncovered the secrets of DNA, spliced genes and achieved many things previously thought impossible. Biomedical science has changed the face of death. Diseases which were once commonplace killers – cholera, typhoid, tuberculosis, diphtheria – have, at least for the wealthy, been all but eradicated. The promise of immortality, or of greatly increased longevity, does not require an especially great leap of the imagination. As Drexler predicted in his early work on nanotechnology:

Future medicine will one day be able to build cells, tissues and organs and to repair damaged tissues...These sorts of advances in technology will enable patients to return to complete health from conditions that have traditionally been regarded as non-living and beyond hope, i.e., dead. (Regis, 1990: 4)

In this way, *fin-de-siecle* hubristic mania expresses the desire for complete sovereignty, perfect knowledge and total power: tired of the ills of that old flesh? Then remake it. Biotechnology signals the ways in which the limitations of the 'natural' body can be fully transcended. New technologies of corporeality hint at the collapse of the temporal distance between the more 'natural' present and the cyborg future, in which bionic bodies enhanced by genetic therapy will be commonplace.

While many of these techniques for extending the body remain potential, the transplant of human tissue and organs into another has become routine – and is again illustrative of social and global inequalities in health. Drugs, such as cyclosporine, which control the body's rejection of foreign objects, became available, transplants became more common and the outcome more certain. However, there are many more people waiting to receive an organ than there are caderveric or live donors; the inevitable result is a booming international black market in body parts and processes that caters to rich recipients and poor or destitute donors. Here the media has had a field day imagining the helpless street children murdered for their corneas – the stuff of modern urban legend – and, while there is little hard evidence of such extreme practices, there is a clear process of commodification emerging, in which body parts can be bought and sold like any other good or service. Indeed, large pharmaceutical companies seeking to develop markets in products derived from human body parts are said to be among the major participants in the trade. Other examples of commodification include surrogacy and egg donation services, which we consider in more detail below, and the very substantial black market in kidneys.

As in other areas where commodification has attended advances in biomedical science, there is an evident North/South dimension to the trade in body parts and processes. The Voluntary Health Association of India estimates that each year more than 2000 people sell their organs for money, many to buyers from developed countries. While the nature of the trade means that prices are extremely variable, we know that it is an exceptionally profitable multi-million dollar business; recipients can be expected to pay up to $50 000, whereas donors can receive as little as $500.[4] In China, according to Human Rights Watch/Asia, 2000–3000 organs a year are taken from the bodies of executed prisoners (www.american.edu/projects/mandala//TED/BODY.HTM). The transplant service is available to high-ranking Chinese and wealthy Hong Kong buyers. Neither legal amendments in India nor threatened trade sanctions against China appear to have had much effect against the trade in human parts. It is ironic that medical technology intended to advance and save human lives has been abused to such a degree. Moreover the contemporary structures of global inequality will ensure that extension of the body's life remains available only to a privileged minority. Accordingly, as Patel points out, moral repugnance may be a luxury of the wealthy:

> It is the culture of poverty that perpetuates the kidney trade in India. In a country where millions suffer from abject poverty, the allure of 'easy money', gained by selling a kidney, becomes an easy trade off for the poverty stricken individual. Given these circumstances, trying to curb the kidney trade may prove to be more detrimental, for India's destitute who struggle to survive can be bettered with the money they can obtain by selling a kidney. It is easy, according to American values and standards, to denounce the sale of kidneys for money as a deplorable and immoral practice but given the fact that the poor in India don't have the liberty of being altruistic because of their harsh existence, the distinction between ethics and survival becomes blurred. (Patel, 1997: 6)

The attempt to exile death, and thus *nature*, from our lives is perhaps the most startling defining feature of modern Western society. While the yearning for immortality can be traced back to ancient civilisations, recent advances in biomedical science – including our knowledge of biological mapping and new medical (micro) technologies – encourage the desire, not only for immortality, but *to live as if death did not exist*. Thus, on the one hand, developments in biomedical science are excessively concerned with the extension of human existence – which is fully commodified. Only the rich can afford the lifestyle associated with longevity and the medical technologies that will keep them in optimum health; and only the extremely poor are desperate enough to sell a kidney on the open market. On the other hand, when we live as if the future never comes, as if we are not mortal, then our ability to connect with nature as embodied and embedded beings is lost. Our traditional knowledge – garnered over centuries of existence – of our bodies and our lands is gradually eroded; our concern for the welfare of future generations lost.

Within an overall critique of the dualisms of modernist society, particularly the sequestering of life from death and nature from culture, feminist philosophers have targeted 'masculinist' science as fundamentally opposed to the (more ethical) concerns of (more 'natural') women. Science's production of disembodied knowledge, impersonal and transcendent, led to the development of the technologies of life and death that caused such anxiety to feminists in the 1970s and 1980s.

The appliance of science

Alongside the feminist critique of scientific method, women have played an important role in actively challenging the application of scientific knowledge. Attention has focused, in particular, upon technologies considered to be threatening to, or destructive of, life. A prominent example, here, is the women's peace movement, which maintained throughout the 1980s, in Europe, the USA and Australia, a high profile campaign against nuclear weapons technology. Also of relevance is the work of Rachel Carson in campaigning against the indiscriminate use of chemical pesticides. In *Silent Spring* (1963), her most influential publication, Carson drew attention to the dangers accruing, both directly to human health and to the ecosystems on which humans depend, from failure to conceptualise the fundamental inter-connections between social (cultural) and ecological (natural) systems. As Carson observes: 'We now wage war on other organisms, turning against them all the terrible armaments of modern chemistry, and we assume a right to push whole species over the brink of extinction' (Carson, 1962, quoted in Hynes, 1989: 7).

Although not writing explicitly from a feminist perspective, Carson's ideas are clearly reflected in later, ecofeminist thinking – not least in relation to developments in biotechnology. Indeed, Patricia Hynes reflects upon the paradox that 'Biology, the science from which Rachel Carson expected eco-logical and life-respecting solutions, is bringing forth its own arrogant and risk-laden technologies' (Hynes, 1989: 177). For later feminist biologists such as Rose (1983) or Haraway (1991) the telos of an unchecked mas-culinist (bio)science was perceived to be biological warfare. While Carson's writing pre-dated these feminist critiques, a decade later she would surely have recognised the continuity of (masculinist) cultural premises which underlie both pest control and population control, and are reflected in the development of biological weapons of mass destruction.

These examples of women's campaigns against life-threatening technolo-gies, in common with feminist critiques of science, reflect an ethic of care and connectedness, which has been associated, implicitly or explicitly, with feminine/maternal principles. Indeed cultural feminists, and in particular ecofeminists, explicitly embrace a maternalist perspective. Arguments, such as those put forward by Chodorow (1978) or Ruddick (1980), formed the basis for most feminist engagements with science in the 1980s.[5] Their align-ment of women with ethical, if emotional, critiques of rational science has persisted. Thus links continue to be made between the patriarchal exploita-tion of the resources of the natural world and the subordination of women and other Others. Both 'women' and 'nature' have been constructed as oppositional to the rational and academic endeavours of a science practised within a globalised system of patriarchal capitalism.

In this way, the portrayal of science as a hegemonic discourse capable of imposing an ideology of 'reality' on non-scientific subjects, and subjugating their knowledges as inadequate, was turned on its head by feminists. Femi-nist epistemology in the 1980s claimed it was actually scientific knowledge that was partial and perverse, lacking the grounded, concrete experiences of those oppressed by its refutation of 'illogic' and the separation of knower from known. For feminist standpoint theorists the situated knowledge of the 'other' was perceived as superior to that of the abstracted and amoral knowledge of the scientist (Harding, 1991; Mies and Shiva, 1993). The power of a (potential) feminist scientific method grew out of this notion of women's embedded and embodied relation to the world. Women as producers of people (subjects) as well as things (objects) had the potential to create a fusion of ontology and epistemology and thus generate more objective, and inclusive, knowledge about the world. In terms of practice, too, many ecofeminists have maintained that women's nurturing capacity should not be confined to the family or community, or even the human species; but that women's ethic of care should be extended to 'nurturing the world'.[6]

The wisdom of assuming disproportionate responsibility for safeguarding the planet has not remained unchallenged by feminists (Bretherton, 1996), and the 1990s have witnessed the emergence of a post-feminist generation of geekgrrls and netnerds eager to embrace rather than challenge the potential of technology.[7] However, in the present context the reactions from *male* defenders of techno-science are of particular interest. Attacks on women activists have explicitly aimed to belittle and control women's behaviour through reassertion of social norms governing the performance of gender roles. Condemnations focused particularly upon the women's failure to conform to traditional models of maternal virtue. Thus, in the case of Rachel Carson, her allegedly solitary lifestyle (she shared her life with female relatives and companions) became a focus of sustained efforts to discredit her work. A recurring theme among critics of *Silent Spring* (including representatives of chemical companies and of the US Federal Pest Control Board) involved reference to Carson as a *spinster* who was more concerned with the death of *cats* from DDT poisoning than the welfare of human beings facing starvation should pesticide use be curtailed. Patricia Hynes' contention (1989: 19) that the treatment of Rachel Carson was analogous to that of 'the witch of Medieval Europe' is, perhaps, not overdrawn.[8] Attacks on the 'peace women' of the 1980s were also virulent, contemptuous and sexualised. They were routinely characterised in the media either as 'unnatural' mothers or as lesbians. As Pettman (1996: 112) has argued, the nature of the invective and the intensity of hostility against women protesters raise important questions concerning 'the power and fury behind assumptions about good women and proper mothers'.[9] More recently, the invective against post-menopausal women who choose to have children by 'artificial' means has been extreme.[10] Indeed these questions concerning the 'proper' roles and responsibilities of women become especially poignant when the complex gender implications of biotechnology are considered.

Reproductive technologies and the meanings of motherhood

Women's uncertainties and ambivalence concerning the potential of new reproductive technologies are associated with fundamental questions concerning the meaning and significance of motherhood. Is motherhood a central facet of women's identity? Is motherhood a right? What is the relative status of biological and social (adoptive) motherhood? Does the desire for children reflect biological imperatives or social pressures to conform?

The availability of new technologies alters the terms of these debates in three ways. Firstly, it extends, in principle, the possibilities for motherhood – not only to women who have experienced difficulties in conceiving naturally but also to single or lesbian women who choose to avoid heterosexual sex, and to post-menopausal or dead women.[11] Secondly, the new technologies have begun to alter the meaning of motherhood. Advances in *in vitro* embryology permit the division of motherhood between genetic mother (whose donated gamete has been used), gestational (or surrogate) mother and social mother, who rears the child. This produces a situation, as John Harris (1992: 1) points out, where motherhood (as has traditionally been the case with paternity) 'is now also merely a hypothesis'. Ultimately, he predicts 'the notion of birth will lose much of its legal and social significance' (*ibid.*). The enthusiasm with which commentators, such as Harris, have welcomed advances in reproductive technologies, including the possibility of exogenesis, suggests an element of male satisfaction which contributes to women's ambivalence concerning their use. The third issue raised by the new technologies also serves to exacerbate anxieties. While women's choice is potentially extended, so too are the possibilities for surveillance and control of women's bodies. In particular, maternal bodies are increasingly monitored – medically, culturally and legally – often at the expense of individual women's bodily sovereignty.[12] Hence the enquiry, 'Are mothers persons?' (Bordo, 1993) is no mere rhetorical device.

At the centre of debates about assisted reproduction is the paradox that, while the new technologies extend the possibilities for attaining motherhood, so the status of motherhood itself is questioned and potentially devalued. In consequence, those feminists who have strongly emphasised women's essential maternalism are also among the most fervent opponents of new reproductive technologies. These critics share wider concerns about women's loss of control of the reproductive process, and of their own bodies, to a patriarchal medical establishment envious of women's procreative power. However their opposition is based, primarily, upon reverence for nature and natural processes, including the generative capacity of women's bodies. Inevitably this strong emphasis upon natural reproduction privileges heterosexual sex as a celebration of nature – 'Reproductive technology alienates both men and women from their bodies and from this most intimate process in which they normally co-operate with their own nature' (Mies, 1993: 139). In addition to implicitly treating homosexuality as unnatural, and thus denying the validity of lesbian motherhood, this idealisation of heterosexual sex contrasts with feminist scholarship concerning patriarchal family relationships, male violence and marital rape. It also appears to ally feminists, on this issue, with an array of religious fundamentalists and right-wing social conservatives.

The apparent contradictions in the position of anti-technology ecological and cultural feminists reflect the complexity of the issues around assisted reproduction. There is an important sense, too, in which this complexity is avoided by a feminist critique of the new technologies which 'makes the hubristic claim to represent *all* women's best interests in utterly eschewing their use' (Farquhar, 1996: 8, emphasis in original). Women have very different needs, interests and aspirations in relation to motherhood, and often it is women themselves who desire access to the new technologies that promise so much. The implications of reproductive technologies are, moreover, differently perceived and experienced by different groups of women. Here women's status, globally – in terms of class, wealth, ethnicity or race – is highly influential in shaping attitudes towards, and access to, the new technologies. This raises important ethical considerations and below we examine two related issues – commodification and genetic manipulation – which have global implications. While our focus on reproductive technologies shows how these are overlaid with gender inequality, a simple association of women with nurturing (or 'nature') is clearly inadequate in a global context where women in the West have access to technologies that oppress others. Nevertheless, women's bodily sovereignty, as compared to men's, has consistently been shown, by feminist scholarship, to be more easily ignored. Thus, in addition to inequalities *across* global locales, women are systematically disadvantaged *within* regions where men and women are supposedly accorded similar legal and social rights (Bordo, 1993).

The commodification of life

The process whereby human parts and processes can increasingly be sold, hired or bought like any other commodity in a global economy has aroused deep misgivings. Human commodification is resisted on the grounds that to own, patent or sell genetic material (or genetically modified organisms) for any purpose is simply wrong. Ultimately this objection reflects the belief that transgenic technology *per se* is an inadmissible interference in the natural processes of life. More specific objections to the procedures involved in new reproductive technologies include the high cost and uncertainty of treatment, payment for donated genetic material and the issue of surrogacy.

Geographically, the availability of assisted reproductive technologies is increasingly global. In the USA, the UK, Europe, Australia, Japan, Israel and the middle-income developing countries, such as India and Mexico, 'treatments' are available for those who can afford them. Cost, however, is a major factor in limiting access. This raises two important issues. Firstly, motherhood itself becomes a function of class, a privilege of the wealthy rather than a right for all women.[13] By extension, in the USA and the UK, it also becomes an issue of race (Steinberg, 1997; Farquhar, 1996).

Financial and other factors that impede access to assisted reproduction are indicative of judgements concerning which social groups should be encouraged to reproduce – and which should not. Secondly, what has been referred to in the USA as the 'stork market' is highly profitable (Keeble, 1994; Farquhar, 1996). The construction of infertility as a medical problem suffered by individual couples, and requiring costly interventions, is big business. In consequence, commodification is expressed through marketing reproductive services in a manner which misrepresents success rates (there are a number of methods of calculating 'success' of which producing 'take-home babies' is only one); through recommending assisted procedures at an earlier stage than necessary, or when the prospects for success are poor; and through generally glossing over the discomfort and stress associated with the procedures, whether they succeed or not.[14] In these circumstances, the centrality of the profit motive involves the exploitation of women/couples who are desperate to conceive and who will be prepared to incur major debts in the process. A further impact of the profitability of assisted reproductive technologies is the diversion of research, and practitioners' efforts, in this field. As in the case of other expensive medical procedures, the emphasis on treatment serves to divert attention, and resources, from investigation of the broader social and environmental causes of infertility; and hence from prevention. Poor nutrition and exposure to environmental hazards, for example, are significantly more likely to affect poor than wealthy communities. It is not incidental that African-Americans are one and a half times more likely to be infertile than their white counterparts (Farquhar, 1996: 42).[15] Scientific and commercial stakes are high, tissues and cells are relatively cheap.

Alongside these broad issues of cost and equity, there are specific aspects of commodification which cause particular anxiety for women. Prominent among these is the issue of payment for donated genetic material. Here we will not address the issues of payment *per se*, which is referred to above. While payment for donation of human tissue (blood) and, indeed, genetic material (sperm) has become widely accepted practice, greater ambivalence attaches to egg donation. This difference reflects the very different procedures, and associated cost differentials, between collecting donated sperm and eggs. In contrast with sperm donation, which is a simple procedure requiring no technical intervention, egg donation involves the donor in a protracted regime of hormonal treatment, monitoring of blood samples and, ultimately, extraction of eggs through invasive procedures requiring local or general anaesthesia. The discomfort and potential risks involved in these procedures raise questions about the motivation of women who volunteer to undergo them; suggesting an exploitative relationship between affluent women who pay for the treatments and poorer women who undergo them. This issue is intensified in the case of surrogacy.

Surrogacy, or gestational motherhood, is an arrangement utilised in circumstances where a woman who wishes to have a child, but is unable or unwilling to bring a pregnancy to term, pays another woman to perform this 'service' for her. Not all surrogacies are commercial arrangements, they may involve family members or close friends, but many are. There are numerous, difficult issues associated with surrogacy, including the possibility of identity problems for the ensuing child. Our concern here, however, is with the potentially exploitative nature of commercial surrogacy arrangements. As practised in the United States, surrogacy typically involves a contract between economically privileged and less-privileged women, involving a substantial fee and additional expenses. For some commentators, this voluntary arrangement between adults is acceptable providing that the payment involved is adequate to compensate for the discomfort and possible risks involved in pregnancy and childbirth (Tong, 1997: 191). However, critics of commercial surrogacy point to the highly unequal relationship between the surrogate and her 'employer'. This can result in very close supervision of the surrogate throughout her pregnancy, not in the interest of her own health but in order to ensure the healthy development of the foetus. Moreover, the decision of poor women to offer their services in this way may be a function of economic necessity rather than free choice. Of relevance here is the preference of some surrogacy agencies for unemployed women or women on welfare, especially single mothers, on the grounds that they are unlikely to decide to keep the baby to which they have given birth, since they cannot afford to forfeit their surrogacy fee or to support an additional child. Indeed the president of one agency (the Bionetics Foundation Inc.) has 'urged the surrogacy industry to move to poverty-stricken parts of the United States, where women are willing to gestate foetuses for one-half the standard fee, or to the Third World, where women are willing to do so for one-tenth the standard fee' (Tong, 1997: 201).

While advances in biotechnology may in the future open up the possibility of successful exogenesis, thus obviating the need for a surrogate, the present application of market principles to reproductive technologies raises fundamental issues of exploitation and inequality. Exploitation of women made vulnerable through desperation – as a result of personal desire, family or social pressure – to conceive a child, and exploitation of other women made vulnerable through poverty. The relationship between these women is simultaneously one of inequality. Broader issues of inequality are raised by these technologies, however. The structural inequalities which underlie unequal access, coupled with the potential implications of genetic testing and genetic engineering, raise uncomfortable questions about the kind of people we want to become.

Genetic engineering and the 'promise' of eugenics

The term eugenics implies selective breeding; the application of notions about the desirable characteristics of human beings. In consequence, eugenic ideas and practices are about 'not only racialised and classed, but also gendered and heterosexist notions of nature, and the heritability of normality and abnormality' (Steinberg, 1997: 33). The eugenicist implications of new reproductive technologies flow from two sources – those that are apparently, or primarily, involuntary; and those that result from deliberate choices or policy decisions.

Some of the 'involuntary' mechanisms producing selective results have already been alluded to in our discussion of commodification. The high cost of assisted reproduction, coupled with poor availability of public funding, produces a class and racial bias in the recipient population. This overarching bias is reinforced by the procedures operated by clinics in selecting recipients. Surveys conducted since the mid 1980s, conducted in the US and the UK, have consistently revealed, among recipients, a preponderance of white, middle-class, able-bodied women living in stable heterosexual relationships. While this may simply reflect the cost factor, there is considerable evidence that physicians routinely make judgements about who is 'childworthy' (Tong, 1997: 166). In her study of all registered clinics in the UK offering IVF and related treatments, Steinberg (1997) found that the selection criteria routinely used were highly subjective. They included intelligence, sexuality and lifestyle, age, financial status, attitudes to treatment and ability to bring up a child. While, in all cases, there was a strong denial that racial criteria were utilised, the racial bias in actual treatments suggests that race is subsumed within other criteria. In the USA, for example, despite the significantly higher incidence of infertility among African-American women, there is a perception of the black female as 'an out of control overproducing breeder' who is, moreover, likely to prove an irresponsible parent (Farquhar, 1996: 74). That is, the black woman neither needs, nor deserves, treatment. It is difficult to demur from Steinberg's conclusion (1997: 33) that IVF selection practices 'can be seen to relate to the reproduction of ableist, class oppressive, (hetero)sexist, and racist social divisions'.

So what of consciously eugenicist practices? These can be divided into two categories – negative eugenics, which relate to prenatal genetic screening and diagnosis and the eradication of undesired characteristics; and positive eugenics, which relate to genetic engineering and the production of desired characteristics.

The technologies of prenatal diagnosis are already widely used. They have recently been enhanced by a procedure that permits the isolation of foetal tissue within the mother's blood, thus obviating the need for amniocentesis. In future, its relative ease and safety may mean that testing for genetic abnormalities, which also reveals the sex of the foetus, will become routine. In consequence, it seems almost inevitable that increasing numbers of women will face decisions – having evident eugenicist connotations – concerning whether or not to terminate their pregnancies. This issue has produced a range of different responses, from the view that 'it is wrong to bring avoidable suffering into the world' (Harris, 1992: 71) to the argument that 'aborting a foetus simply because it has Down's syndrome sends a very negative message to actual persons with Down's syndrome' (Tong, 1997: 235). For feminists this issue is particularly difficult, both because the burden of caring for a severely handicapped child is likely to fall disproportionately upon its mother, and because feminists have strongly opposed abortion on the grounds of the sex of the foetus.

This issue has broad ramifications. It must be viewed in the context of an analysis that regards population growth in the South as a major problem threatening the future survival of the planet and which has resulted in concerted efforts, over the past twenty years, to control the fertility of Third World women. Even before the technology was available, the notion of the breeding male was proposed as an ideal solution to the 'problem' of excessive fertility by a group of biologists. Thus, it was claimed, 'Countless millions of people would leap at the opportunity to breed male (particularly in the third world) and no compulsion or even propaganda would be needed to encourage this, only evidence of success by example' (quoted in Mies, 1986: 124). In countries such as India and China, where proactive population control programmes are in effect and preference for male children is strong, there is considerable evidence of disproportionate termination of female foetuses.

It is evident that current utilisation of prenatal diagnostic procedures shows distinct patterns of North/South difference. In the North they are 'primarily associated with middle-class demographic needs and desires – smaller families, better babies, etc.' (Farquhar, 1996: 173). In the South they are used as an instrument of population control through selective breeding of male children. In both cases important judgements are evident concerning those members of society who are valued and those who are not. Although these evaluations are social constructions, that is inequalities are not *caused* by new technologies but reflect the value systems of the societies that produce them, the availability and use of these technologies serves to reaffirm social prejudices – and to obviate the need to seek alternative approaches to complex social issues.

While genetic screening procedures are likely to become increasingly sophisticated, and raise ever-more perplexing moral dilemmas, the basic techniques are currently available. Thus contemporary practice is increasingly indicative of negative eugenics. However positive eugenics – deriving from genetic manipulation of human embryos to produce desired physical characteristics, enhance intelligence or influence personality traits – is relatively in its infancy. It is claimed, nevertheless, that it will become 'routine practice' within the next twenty-five to thirty-five years (*The Guardian*, G2, 14 April 1999).

Advances in biotechnology mean that beef and dairy farmers can already pre-select the sex of their calves by a method which sorts the bull's sperm into two groups, those with X chromosomes and those with Y. It is likely that similar techniques, although expensive, will soon be available to humans. This technology will fuel the ongoing ethical debate, not only about how far we are justified in interfering with our genes generally, but whether the use of technology for reasons considered trivial is ever acceptable.[16] As suggested above, such technologies are likely to be limited to the wealthy, and we run the risk of creating an inherited elite society that increases rather than decreases the poverty gap, exacerbates class distinctions and racial and sexual tensions and significantly devalues disabled people. Here an interesting precedent for the type of choices which could be made is the attempt to produce 'superbabies' by the Repository for Germinal Choice, established in California in the 1980s, where women paid to be inseminated with sperm from Nobel Prize winners.

While the potential to breed 'better babies' has become a matter of public (and media) anxiety, it is the broader political implications of genetic engineering which raise greatest concern. Clearly, alarmist predictions about the ability of science to create a master race greatly anticipate the scientific possibilities. Nevertheless, there is a long tradition, from Plato to the Third Reich and beyond, of theorising and state practice intended to exercise 'quality control' of populations. The most highly developed, and explicitly articulated eugenicist policy of recent years was introduced by the Singaporean government in the 1980s. This involved a range of financial and other incentives for graduate women to bear children, and was accompanied by payment of substantial rewards to less well-educated women who kept their families small (Chan, 1992). Similar, albeit less explicit, eugenicist policies were widely reported by delegates to the 1994 United Nations Conference on Population and Development. Of particular significance, in the 1990s, has been the strongly eugenicist rhetoric accompanying the resurgence of ethno-nationalist forms of political mobilisation. This has been evident in the Balkans and much of Central and Eastern Europe, where women have been exhorted to participate in a 'demographic race' with neighbouring ethnic groups (Yuval-Davis, 1997: 30). This emphasis upon women's duty to procreate in the interests of the nation contrasts with the strident population control discourses which, elsewhere, problematise women's fertility. In both cases, however, it is state control of women's choices, and women's bodies, which is at issue.

These examples illustrate the recurrent eagerness of state governments to regulate the quality, and the quantity, of the 'national' population. It can only be assumed that governments will, in the future, attempt to exploit the potential of genetic engineering to this end. Thus, again, new reproductive technologies suggest, simultaneously, the prospect of increased social control and the opportunity for enhanced individual choice. While causing anxiety, they also tantalise – for 'If it is not wrong to hope for a bouncing, brown-eyed, curly haired and bonny baby, can it be wrong deliberately to ensure that one has such a baby?' (Harris, 1992: 161).

'These are the days of miracle and wonder...'

Medicine is magical and magical is art
The boy in the bubble
And the baby with the baboon heart.

(Paul Simon 'The boy in the bubble')

In this chapter we have considered women's uneasy and ambivalent relationship women have with the development and application of scientific processes. Areas of traditional significance to women, such as child rearing, have informed feminist critiques of transcendent science and have been the focus of our analysis here. We have argued that, in the information-rich and wealthy North, highly expensive procedures that widen reproductive choices are available as a solution to the 'problem' of infertility, although the availability of these procedures is significantly influenced by subjective assessments of individual women's suitability to parent. Furthermore, one's wealth and status similarly influence the availability of life-enhancing medical treatments. In the South, on the other hand (or for other less-worthy recipients of medical wonders), the 'problem' of excessive fertility has merited (male/Western) intervention in the name of population control. At the same time impoverished countries, with so many 'spare' people, are developing a black market in body parts that services the wealthy. Thus the biogenetic replication and regulation of our bodies is fraught with ambivalence, while the spectre of eugenicism cannot be ignored.

The overarching condemnation of radical ecofeminists and the morally conservative fails to address the desires of the many women (and others) who are desperate for the promises – of fertility, healthy crops, a new lease of life – that techno-science holds out. And it is often women who stand to gain the most from safer births and healthier crops and children. On the other hand, we are genuinely fearful of the consequences this God-playing arrogance might bring: the baby with the baboon heart, the Elvis clone and the trade in body parts are still seen as 'unnatural' wonders. While post-humanists embrace the encroaching cyborg future (see Pepperall, 1995) many women are uneasy about the propriety of techniques which derive from experiments with aborted foetal matter or that aim to create and modify life itself.

A modernist, dualistic science that sets itself apart from women and nature is unable to reassure. We have argued that a renewed and empathic relationship with nature and respect for other, previously ignored knowledges, could generate a more ethically grounded perspective on biotechnology. As living organisms embodied and embedded within local and global ecosystems, we require a more reflexive analysis of the impact biotechnology has on a nature reconceived (like women) as the *subject*, rather than mere object, of scientific knowledge. An analysis of biotechnology as a global gender issue, fundamentally altering pre-existing patterns of production and reproduction, has shown how inseparable from culture nature has become.

Notes

1 The notion that death is not a part of life, an ongoing and inevitable process, is a particularly Western belief; a view that emerged in the Enlightenment period when science attributed causes to events. See Nuland (1994) and Pepperall (1995).

2 Moreover, in the West there is an increasingly moral dimension to health and illness that influences medical and social responses to particular individuals. HIV/AIDS rhetoric has long distinguished between 'innocent victims' and those whose practices mark them as culpable; smokers have been denied expensive surgery because their habit signifies a lack of commitment to good health.

3 The low-tech responses to spiralling heart disease figures include increased sales of health foods and all those middle-aged runners in our parks. New body technologies, such as these are promoted as life saving as well as life enhancing, and medical discourses promote a self-conscious self-surveillance in which we regulate our consumption of nicotine, alcohol, fat, sunlight, exercise, salt, sugar, cholesterol and so on. Only our determination now will delay the (currently) inevitable.

4 Inter Press Service, http://www.oneworld.org/ips2/oct98/09_25_024.html. Other useful resources can be found on the Trade and Environment Department and the United Network for Organ Sharing home pages.

5 These writers argued that the differing epistemologies of men (as abstract knowers) and women (as more ethically aware) were related to their different ontological experiences as children. While girls learnt their future role from a concrete and caring mother, boys, in their separation from the feminine sphere, had only an abstract (and absent) entity with which to identify.

6 'Women Nurture the World' was the title of a workshop in Nairobi, in 1985, organised by the United Nations Environment Programme.

7 See Terry and Calvert (1997) for an account of a techno-friendly cyberfeminism that explicitly rejects the earlier association of women with nature and with technophobia. Web sites such as geekgrrl ('grrls need modems!') are also dismissive of old-style 'victim' feminism suggesting that via our modems 'grrls' can begin to transcend the limitations of bodily based idenities (www.geekgirl.com). The alternative spellings that abound serve a practical purpose to subvert the easy appropriation of women's space. As Crystal Kile (creator of the PopTart pages) says 'grrls/geeks/nerds use these codewords in titles of our site to make it clear that we're not naked and waiting for a hot chat with you!' (*ibid.*: 60).

8 In addition to the explicitly gendered attacks upon Carson, *Silent Spring* was also depicted as part of a Communist plot intended to undermine US agricultural and industrial production. A number of chemical companies attempted, through threats of legal action, to prevent publication of *Silent Spring*, while immediately before its publication the (then) Monsanto Chemical Company issued a parody which it had commissioned entitled 'The desolate year'. Corporate funding was subsequently withdrawn from broadcasting companies that conducted interviews with Carson (Hynes, 1989: 17).

9 In a similar way, geneticist Barbara McClintock, who disputed the accepted scientific model as objectifying, was dismissed by her male peers as 'difficult' and professionally marginalised (Mellor, 1997: 118–19).

10 It should be noted that the male doctors who help women achieve such an unnatural state of affairs have not been immune to critique. For example, *Daily Mail* (29 April 1999) described Dr Severino Antinori, who has been involved in helping the world's oldest mothers to conceive and has promised fourteen women a millenium baby (the caesareans start seconds after midnight), as the 'world's most dangerous doctor'.

11 There have been a number of high profile cases in which the bodies of brain dead women have been kept alive artificially, sometimes for many weeks, to allow the foetus to grow and develop further.

12 Not only have pregnant women been denied the right to abortion or kept 'alive' against their wishes, but a number of pregnant women in the USA have been imprisoned, on the charge of child abuse, for using alcohol or drugs.

13 Here it is of interest to note that, whereas the medical profession and society increasingly construct 'infertility' more broadly, as a disease requiring treatment, it is not so regarded by most public or private health service providers. Only in France can women reclaim the cost of treatment from the state. In the UK, despite the recommendation of the 1984 Warnock Committee that infertility is 'a condition meriting treatment', 90 per cent of treatment takes place outside the National Health Service (Keeble, 1994: 11–12). This distinguishes assisted reproductive technologies from other costly medical procedures – particularly Viagra which is available on the NHS.

14 In the UK the Human Fertilisation and Embryology Authority, which was established in 1991, regulate these practices to some extent. However they are widely reported in the USA and continental Europe (Farquhar 1996: 78–9). See also the figures provided by the *Center for Surrogate Parenting* (CSP) which stress the number of couples 'at home with babies' while downplaying both the possibility of complications and the cost of paying for them. Moreover, as we note below, the CSP service, estimated to cost at least $50 000, is clearly limited to very wealthy couples.

15 While the financial dealing of eggs, sperm and wombs is reasonably well documented, the trading of organs as commodities on the open market is less well known. However, what information we do have suggests that the impoverished, particularly women who have lower social status, are most at risk of exploitation. Thus the 'voluntary' sale of kidneys in India is disproportionately borne by women.

16 Alongside these moral dilemmas, more practical matters also require careful consideration. Our desire for the future encourages us to rush through the implementation of technologies when their outcome is unpredictable. Despite thalidomide, despite BSE, we too often forget that gene manipulation might well cause unpredicted congenital handicap or disability; gene therapy is extremely complex and the outcome is currently uncertain.

Part II

International political economy, trade and environment

Taken together, the chapters in part II are sufficiently complementary to represent a general perspective or orientation on the significance of biotechnology for IPE. The chapters collectively address the commercial portent of biotechnology and the international consequences of a rush to exploit its diverse potential. Williams (chapter 5) opens the discussion by explaining the importance of the Agreement on Trade Related Aspects of Intellectual Property (TRIPs) – negotiated under the Uruguay Round – for knowledge-intensive biotechnology. His chapter identifies the centrality of intellectual property for an economic analysis of a knowledge-based generic technology. Biotechnology firms are shown to have campaigned successfully to achieve extension of the international patents regime into areas of the exploitation of genetically modified organisms and biological production processes. Consequences of this development extend to the control of the seed market, biodiversity and the general structural characteristics of the IPE. In this respect, structural characteristics of the global economy are seen in the fashion by which firms, with the support of governments of key industrial countries, have an ability to wield structural power – what Strange has described as the ability to shape global structures (1988, 1994) – to achieve a global playing field of their design.

Russell (chapter 6), explores, at a more conceptual level, how biotechnology might be located within the knowledge structure of the global political economy. He paints some alternative ways by which biotechnology, as technology, can be internalised into the concerns of the field of International Political Economy. Russell also addresses the generic nature of biotechnology and the many trajectories it can follow within a common technological paradigm. The chapter goes on to examine a particular perspective described as Actor-Network Theory (ANT), which opens up the debate over the networking of many types of actor, a focus which Pownall elaborates on, although from a less radical stance than ANT. ANT is notable for the way it incorporates technology and its artifacts within networks, cautioning against restricting network discussions to human agency at the expense of externalising technology.

Pownall (chapter 7) follows both Williams and Russell in recognising the significance of transnational networks of firms. He utilises a structural perspective (Strange, 1988, 1994; Stopford and Strange, 1991) where structure is again reflected in the pattern of inter-firm and firm–government alliances. This is then developed in great detail by looking at the regional strength of Europe in biotechnology-related industries – in some respects challenging the structural power of the United States. Pownall also reinforces the need for an interdisciplinary analysis of biotechnology, bringing aspects of IPE, business studies and policy analysis together. There is a contrast here with Russell's use of parts of the sociology of technology.

While the first three chapters in this part take account of the controversial character of biotechnology, something that has given rise to a disparate set of actors and issues involved, one very important and long-standing concern is drawn out by Vogler and McGraw (chapter 8). This is the international requirement to address the safety issue. From its inception new biotechnology has raised concerns over safety: in the 1970s and early 1980s it was laboratory safety; by the 1990s it had become industrial scale-up and environmental release. These latter concerns reached new heights when combined with an environmental focus on biodiversity. Biodiversity is itself another controversial issue raising questions of ownership, control of genetic resources and their exploitation. Williams establishes the importance of rights over genetic resources and this complements the focus on the environment in the chapter by Vogler and McGraw. An important aspect of biotechnology is its relationship with biological raw materials and their manipulation. With all biological organisms in turn featuring in what we describe as the natural environment, we cannot ignore the environmental impacts of biotechnology. These extend from resource depletion (as GMOs come to displace natural varieties) to environmental release (with possible associated hazards). This chapter reviews efforts to establish and operate an international environmental regime for biotechnology. The key question is whether there will be a regime which emphasises trade or which adopts a precautionary approach to the protection of the environment – under the Biosafety Protocol of the Convention on Biological Diversity. At stake is a global policy response to the protection of the environment from the release of living modified organisms, in a charged political climate of diverse and competing interests.

The seed market was an early focus of attention for patenting activity, given the potential to re-sow future generations of a proprietary seed. Underpinning this is the development of genetically modified foods. The global political economy is beginning to witness a shift towards restructuring of agriculture as new modified seeds come, perhaps rather forcibly, to market, as large and powerful firms push hard in this direction. Williams illustrates some aspects of this power in the lobbying for TRIPs, while Pownall observes overall industry characteristics in Europe. GM food has many features, including seeds and GM techniques applied elsewhere in the food production process. Falkner (chapter 9), rounds off this section, by analysing the trade consequences of biotechnology developments in the global food industry. His chapter brings together a number of themes established in earlier chapters: noting the disparate range of actors from firms to environmental pressure groups; European and United States differences in perception over the progress of biotechnology, industrial and agricultural exploitation; differences which in turn are a basis for trade disputes and struggling institutional responses from seeking a biosafety protocol to managing dispute settlement; and Falkner summarises overall differences in North–South perspectives.

5

Life patents, TRIPs and the international political economy of biotechnology

Owain Williams

The relationship between biotechnology, intellectual property rights (IPRs), and production based on the exploitation of biological resources constitutes one of the most challenging and nebulous issue areas of international political economy today. Certainly, the implications of the patentability of life forms have already become closely associated with the emergence of biotechnology as one of the most important technological developments of the twentieth century, and are likely to produce serious conflict in the global political economy of the next century. This is largely because biotechnology allows for the manipulation of the most basic requirements of human life – food, materials and medicines – and is steadily changing the manner in which these goods are produced at the most fundamental level.

This chapter seeks to flesh out the links between biotechnology and the new intellectual property rules for life forms that have been incorporated into the World Trade Organisation. Both the agreement and the growth in rights over genetic resources are viewed as part of a movement towards increased corporate control over technology and broader global knowledge structure (see below). In particular, the agreement in Trade Related Aspects of Intellectual Property – or TRIPs – has profound consequences for global agricultural production and the livelihoods of people in the developing world. The analysis of the agreement concludes with an assessment of the implications of 'strong' IPRs for the development of biotechnology itself – especially with regard to the growing availability of life patents and Plant Breeders Rights on a global basis.

Patents and biotechnology

The political economy of biotechnology, as one of the most important technology systems of the twentieth century, has been closely associated with the increasing availability of IPRs in the field. Indeed, since the first patent was awarded for a modified micro-organism in the United States in 1980, commercial interests in the technology have skyrocketed. The United States' lead in life patenting has been mirrored throughout the states of the developed world, with incrementally stronger and more commercially oriented protection becoming available in the European Community and Japan (most recently witnessed with the 1998 adoption of the European Patent Directive). The rise of life patenting has seen the adaptation of classical intellectual property law to accommodate a completely new science, and the marriage has never been a happy one. Legal challenges to life patents have largely been based on the morality of giving any one single agency what amounts to exclusive rights to a particular life form or genetic characteristic. Critics of biotechnology patents have also pointed to the injustice of allowing patent rights for 'inventions' which already exist in nature. These types of patent are often solely awarded on the basis that a particular gene has been isolated for a given productive purpose, or has simply been added to an existing life form (as is the case with many bio-engineered plants). Indeed, it is often argued that genes and living organisms are 'products of nature', and even when isolated, altered or purified, cannot be classified as inventions as they are purely 'discoveries'.

Opposition to life patenting in the developed world shows no sign of respite, but the use of IPRs by industry has become more widespread nonetheless. Patents perform a very important role in the political economy of biotechnology in the developed world. In the main, they have principally allowed a system within which commercial monopolies can be re-enforced over the technology and across the economic sectors which the technology has impacted upon (see below). However, in terms of biotechnology knowledge, production and consumption, life patents have also supplied a number of other functions *vis-à-vis* biotechnological knowledge control.

First, and as is intimated above, many of the most basic developments in biotechnology have emerged from the universities and public research institutions of developed countries – and this includes the breakthroughs of recombinant DNA and monoclonal antibodies (see chapter 2 by Manning in this volume). State funding has been instrumental in bearing a large proportion of the huge R&D costs and basic research undertaken in the field. Companies have also been given tax breaks and other incentives to effectively commercialise upon 'publicly' generated knowledge. A pattern of state sponsored industrial development of biotechnology has been particularly evident in Japan and the European Community (Pownall, chapter 7, this volume). Patents have, therefore, supplied an important and essential means by which private and exclusive biotechnological knowledge can be demarcated from the broader milieu of knowledge production in the field. The cumulative and cross-disciplinary nature of the scientific knowledge involved in biotechnology often means that the boundaries between research undertaken, say, by universities, and that used productively by TNCs, are potentially too blurred as to justify exclusive exploitation by private corporations alone. Patents are therefore used as a type of legal boundary between the overlapping private and public domains of knowledge – and have provided a key mechanism by which knowledge is transferred from academia to the corporation (Kloppenburg, 1988: 196).

Second, basic research in biotechnology is often the source of new processes which can provide a scientific starting point (or means of production) to activities in different economic sectors. But the knowledge involved in biotechnology is also characterised by the inapplicability of divisions between 'basic' and 'applied' research, to such an extent that 'the basic scientist on the campus doing basic research may end up doing some extraordinarily important industrial research' (Harsanyi, 1981: 181). Patents are therefore important for controlling the underlying scientific bases of the technology and, again, for demarcating the public domain of knowledge from the private and exclusive. The ability to patent a new development that has emerged from basic research – such as a biotechnological process – is thus extremely important, and supplies an enormous degree of commercial control over the productive activities that the process is used in. Since corporations have only played any meaningful role in basic biotechnology research in the last decade (see in general Kenney, 1986; George, 1990; and Kloppenburg, 1998), the availability of patents for products and processes which are derived from publicly generated knowledge express an instance of knowledge control by means of legal fiat. For George, the appropriation of biotechnological knowledge by TNCs represents a somewhat startling assertion of corporate power given the historical evolution of the field:

> Today's biotechnology came from the work of thousands of people who patiently dug the foundations, built the walls and raised the roof-beams of an enormous edifice. These prodigious labours now accomplished, corporations new and old are jostling one another on the building site to put the final slates on the roof and call the whole place their own. (George, 1990: 115–16)

Third, the patent and patenting licensing system has also facilitated a form of contractual framework within which transnational corporations have managed their commercial relations with universities and small biotechnology firms. TNCs have increasingly invested in their in-house research capabilities, but have also continued to develop new products from out-sourced research (Pownall, chapter 7, this volume). Maintaining a raft of licensing agreements has allowed them to effectively stay on top of a complex set of knowledge systems and knowledge producers, and the power to cherry-pick the most promising research lines.

However, the most immediately apparent feature of the socio-economic relationship between patents and the emerging international political economy of biotechnology lies in the manner in which the state-sponsored patent system has led to increasing commercial dominance of a number of key sectors of production in the global economy. This has resulted from the specific advantages that a patented product or process grants companies in terms of competitive advantages (such as barriers to entry into competition) and from the technological protectionism that patent rights provide across global markets. The availability of strong patent rights in the field has promoted unprecedented corporate control of the life sciences.

Patents are, after all, temporary monopolies – giving the patentee the right to exclusively use, make, or sell a given invention. These rights have historically been exercised over a defined national market for a fixed period of time (ranging from 10–20 years). Patents have thus, ostensibly, supplied a form of politically sanctioned guaranteed reward for inventors, in which further inventive activity has been fostered. However, the right to reward has always been balanced by the broader public interest – and IPR systems have been fashioned to reflect divergent national economic goals or technology policies (Williams, 1998). In more general terms, these political limitations have meant that national patent laws have certainly never been subject to a uniform set of rules or, indeed, to a globalised legal regime. Where patents have been available, they have almost invariably offered protection only if an invention is deemed as meeting a number of criteria. To qualify for a patent, an invention must represent an inventive step (or be new), be non-obvious (to another expert in a given field), and be capable of industrial application. Patents are thus particularly suited to industrialised systems of knowledge production. But, globalisation of production and markets has led corporations to seek a commensurable legally uniform regime in which to manage their interests. This regime has been imposed on many countries regardless of the level of economic development or the structural characteristics of their modes of production.

TRIPs and patents

The Agreement on Trade Related Aspects of Intellectual Property, or TRIPs, is one of a triad of agreements (including the Agreement on Services and the new GATT) which are administered by the World Trade Organisation. The agreement which came in to force in 1995 was one of the most contentious and divisive of all the negotiating areas of the Uruguay Round of GATT negotiations (1986–93).

TRIPs details rights and obligation in seven areas of intellectual property (patents, copyrights, trademarks, industrial designs, chip layouts, geographical indicators and trade secrets) which must be enacted by all parties to the WTO. The regime represents an attempt to provide harmonised 'minimum standards' of IPRs in the international political economy, and effectively supplants preceding international conventions in this area (such as the Paris and Bern Conventions). The use of the term minimum standards is often misleading in the context of TRIPs, as the agreement is generally recognised as strengthening the IPR protection of technologies as compared to that available under the major conventions (see variously Barton, 1995a; Reichman, 1995; and Correa, 1992). TRIPs will also force many countries to provide IPRs for technologies that they previously excluded from their national systems – and most notably it forces the majority of developing countries to protect pharmaceutical, agrochemicals and plant varieties for the first time (Braga, 1996). This is not surprising, as TRIPs is very much an extension of the IPR jurisprudence of developed and industrialised countries to the national legal systems of their developing counterparts.

TRIPs is an extremely intrusive regime, giving WTO members little-or-no scope to tailor their IPR policies for reasons of national economic or developmental needs (De Almeidia, 1995; Barbosa, 1995; and Chimni, 1993). WTO members cannot simply pick and mix particular types of IPRs that they feel are appropriate to their national systems. Life patenting is thus a compulsory element of the job lot of obligations that make up TRIPs. Moreover, countries could not merely sign to individual WTO agreements of their choice, as the WTO is also an 'all or nothing' package. To make the provision of IPRs doubly difficult to evade, TRIPs links the enforcement of IPRs in member countries with the WTO dispute settlement machinery. This body was substantially revamped and given teeth in the Uruguay Round. If a state fails to enforce the agreement, or protect the IPRs of foreign nationals, then cross-sectoral retaliation (or economic sanctions) is permitted against the offending state. The beefed-up dispute settlement machinery of the WTO is already beginning to flex its muscles over the rules under which global trade is conducted, and, if sufficiently strong to force substantial upheaval in trade relations between the EC and USA (as has been the case in the 1999 banana and beef growth hormone disputes), it will doubtless impact more on countries with smaller trade weights than these parties. In general terms, countries have a single choice in terms of IPR and life-patenting laws – namely to accept the rules or defect from the WTO. This, in essence, is no choice at all for any single developing country.

These new rules have attracted sustained criticism from many in the South who believe that the regime will entail more costs than benefits to the world's poorest countries. TRIPs has been judged as employing a 'one size fits all' approach to IPRs that fails to reflect the marked disparities between the economies of WTO member states (Gaia/Grain, 1998b: 2). Others have predicted that patents will likely widen the gap between global technology rich and poor (Bifani, 1989; Chimni, 1993; Barbosa, 1995). Moreover, since the South is especially dependent on production which is based on the exploitation of biological resources for agriculture production (three quarters of India's workforce, which has the largest population of any WTO member, is engaged in this area), any rules which gave rights over these resources was felt to be a direct attack on livelihoods of people in the South. It is for this reason that the rules pertaining to the patent protection of life forms and biotechnological products and processes were the most contentious area of the TRIPs negotiating group, and the stimulus for 'trade-related' issues spilling over into the CBD and other international fora.

Life patenting

Article 27 of TRIPs forms the central plank of the regime's rules for patenting (GATT, 1994). The text of the article is so extremely nuanced and technically complex that even a full exposition of the single sub-paragraph relating to biotechnology is beyond the scope of this chapter (for more detailed discussions of the article see variously Correa, 1992 and 1994). However, Article 27 supplies what have been termed Paris Convention–plus rules (Reichman, 1995), since the agreement represents a radical strengthening of the patent requirements of the only existing international patent treaty. IPRs and patents are also defined as purely 'private rights' in the preamble to the TRIPs. This represents a subtle but nonetheless radical shift away from the traditional 'balance of rights' between private inventors and a given broader public/national interests, a balance which was historically present in both national and international IPR systems (see Preamble to the TRIPs agreement and Williams, 1998a). Defining IPRs as private rights also underscores the fact that by enforcing TRIPs many states will be protecting the interests of foreign firms irrespective of the policy preferences of their citizens.

Article 27.1 of TRIPs provides 'catch-all' patent protection for any technology or process for making things. Article 27.2 allows countries to exempt specific inventions from their patent systems if they might conflict with public order, morality, or constitute serious prejudice to the environment. These exemptions would, therefore, seem to offer enormous scope to states wishing to evade providing patents for plant varieties or modified animals. However, the exemptions are generally held to be of a purely cosmetic nature. A senior member of the WTO secretariat has already intimated to this author that any attempt to avoid biotech patenting on these grounds will be swiftly dealt with by the dispute machinery. Moreover, many of the challenges to life patenting in Europe in the 1980s were often fought out over the morality of the private control of any life form (whether modified or not), with the patentee usually winning out on grounds of the application of human intervention to an organism and/or the resulting novelty of the organism concerned.

Article 27.3(b) is the sole section of the TRIPs agreement which sets out the rules covering biological resources (including plants animals and microorganisms) and biotechnology. It states:

> Members may also exclude from patentability: plants and animals other than micro-organisms, and essentially biological processes for the production of plants and animals other than non-biological and micro-biological processes. However, Members shall provide for the protection of plant varieties either by patents or by an effective sui generis system.

At a first reading of the text it would seem that countries could exercise an enormous amount of discretion as to whether or not they permit life patents. The sub-paragraph allows exemptions to the patenting of 'plants and animals', as well as 'essentially biological processes'. Part of the problem involved in understanding the real meaning of the sub-paragraph results from the vain attempt to mesh intellectual property law with a new technology, and it is not surprising that the article has been criticised for its technical inadequacies (Barton, 1995a). However, it is clear that the term 'essentially biological processes' simply refers to the non-patentability of traditional breeding methods – and not biotechnological processes. This means that TRIPs gives no rights to those producing plants by means of selectivity and cross breeding, but insists that the patenting of plant varieties and animals be permitted if they have been bio-engineered or micro-biological processes have been employed. These biotechnological methods are, per se, not understood to be 'traditional methods' of breeding plants – or, indeed, as being biological processes.

The real power of the article lies in the stipulation that countries allow for the patentability of 'micro-organisms' and 'micro-biological processes'. These two elements of the article's patent provisions will alone give rise to a strong and breathtakingly broad scope for international patent rights over biotechnological products and processes (including modified plants, animals and lower organisms). In the first place, micro-organisms (such as bacteria and viruses) provide biotechnology with the major means of culturing, cloning and delivering genetic information. Many developments in biotechnology are likely in some way to be based upon the use of an engineered micro-organism. They are also often products in their own right. It is equally true that micro-organisms are of central importance to the field, since they are a key source of the restriction enzymes which are essential tools employed in isolating desirable genes for transfer.

Similarly, the patentability of micro-biological processes allows an equally broad scope for biotechnology patenting. The term essentially describes any of the characteristic techniques and processes of biotechnology, since the technology after all operates at the micro-biological level. Taken together the patentability of micro-organisms and micro-biological processes allows for rights that are both scientifically and commercially broad. Corporations will, therefore, be able to exercise the same sweeping rights internationally (that is, in the 130 plus member countries of the WTO) as they have enjoyed under-developed country patenting regimes.

The sui generis system and sustainable agriculture

Many critics of Article 27.3(b) feared that the most catastrophic consequences of the sub-paragraph would emerge from the stipulation to provide protection of plant varieties by means of patents, or by adoption of a sui generis system (simply meaning a unique form of IPRs). Plant Breeders Rights (PBRs) have been used throughout the twentieth century to protect varieties of plants, which have been bred for the agricultural systems of industrialised countries. To qualify for plant variety protection the plant variety had to meet the criteria of genetic distinctness (from other varieties), uniformity, and stability (through successive generations). PBRs are thus solely available for crops and plant varieties which are genetically uniform (such as monocultures and hybrids), and not for diverse crops bred to meet local environmental conditions and nutritional demands. This is one reason why the Union for the Protection of Plant Varieties (commonly called UPOV after its French derivation) only managed to attract twenty-five states from its inception in 1961 to the completion of the Uruguay Round. PBRs are simply irrelevant to the needs of farmers and plant breeders in the South.

Despite the fact that patents and PBRs simply do not serve the agricultural needs of the South, the WTO has made it very clear that they see UPOV as the 'off-the-peg' benchmark for meeting the TRIPs requirement that an 'effective system' of PBRs be established. UPOV was actually strengthened in a 1991 revision of the treaty, mainly as a response to the increasing desertion of the PBR system by biotech and seed companies in favour of the more stringent protection afforded by plant patents. The revisions to the treaty have, in the main, further infringed upon the rights of farmers to re-use the seeds of protected varieties for the purposes of re-planting and inveigh upon the right to breed new varieties of plants from protected varieties. In the meantime, UPOV is promoting itself as the sui generis system of PBR in a series of regional meetings throughout the developing world. In the run up to the year 2000 deadline for TRIPs compliance (for developing country members of the WTO), more and more countries rushed to accede to the agreement, fearing trade sanctions if they do not have a system in place.

Whatever system of sui generis rights developing countries eventually opt for, it is fairly clear that PBRs and patents provide a legal incentive to breed uniformity. It is also significant that IPRs will restrict the rights of farmers and communities to use biodiversity. Free access to seed is a key issue in terms of the global agricultural system and global food security – as the seed is the most basic means of production for developing world farmers. Free exchange of seed between farmers is also the most important means by which diversity is diffused and ensured. The jointness and non-exclusivity of traditional knowledge systems are also necessary adjuncts in sustaining food security at both the local and global levels. Ostensibly, patents impose rights over an industrially oriented knowledge system and would therefore seem to have no bearing on traditional knowledge, or the right of Southern farmers to save and exchange seed. This is not the case, as the distinction between traditional and industrially oriented knowledge is more often than not a purely abstract one when the contribution of the Southern agri-biodiversity to the end products of plant biotechnology (or indeed modern plant breeding in general) is viewed over the longer term.

If seed corporations could claim that their engineered varieties and monocultures owed no debt to knowledge systems that characterise sustainable agriculture, then there would be a small case in favour of rewarding their efforts by IPR protection. However, this is not the case. The vast majority of arable and general food crops used in production in industrialised countries are derived from varieties of plants which have been developed over millennia by third world farmers (Vavilov, 1951; Juma, 1989; Kloppenburg, 1988; McDougall and Hall, 1996). Patents will also play a parallel and equally unjust role in the appropriation of ethnobotanical knowledge and resources of the South. As is the case with modified plant varieties, the presence of patents for ethnobotanical resources serves to make the livelihoods and knowledge systems of the South invisible. At stake is not only the future sustainability of traditional knowledge, which can often be lost in a generation if strictures on its use and diffusion are enforced, but the vast profits that appropriation confers on the patentee. The estimated annual value of medicinal plants used by Northern companies for pharmaceuticals (often secured in developing countries by specialised biopiracy firms) is $32 billion, and the value of 'undiscovered' Southern plant-based pharmaceuticals in tropical forests alone is placed at roughly $150 billion (RAFI, 1994). These resources eventually become patented products that can be re-sold to their country of origin.

Controlling access to seed by means of IPRs will open up a means of securing even vaster corporate profits than is currently the case in the South to North transfer of medicinal plants. Seeds are, after all, the fundamental means of production for farmers and agriculture is the single biggest economic sector world-wide (Kloppenburg, 1988; Juma, 1989). A subsequent rise in global seed prices is only one of the concerns of critics of the TRIPs rules. Intellectual property rules for plant varieties are also likely to lead to a net loss in the diversity of knowledge systems surrounding plant breeding – and this will lead to a commensurate loss in global biodiversity. There is simply no evidence to suggest that the presence of globalised plant variety protection will benefit the environment, the diffusion of knowledge, or the ultimate producers of food. Studies on the impact of PBRs on plant breeding in the United States have concluded that the system fails to significantly stimulate the development of new varieties and has led to a net reduction in the flow of information and germplasm from the private to public sector. Of course, seed prices have also risen in that country (Gaia/Grain, 1998b; Butler and Marion, 1985; Buttler, 1996).

TRIPs rules for plant variety protection spearhead an attempt to assert corporate monopoly control over international agriculture by supplying private rights for scientific knowledge as applied to plants (and to animals). IPRs can be claimed irrespective of the contribution of traditional knowledge to the global agri-biodiversity. The continued plundering of these resources for the purposes of generating new and genetically uniform plants plainly highlights the injustices of the international flow of plant germplasm (Kloppenburg, 1988: 15).

The international availability of IPRs for plant varieties adds a new dimension to the asymmetric international transfer of biological resources. Southern biological resources can be obtained at little or no cost to the plunderer, and are increasingly patented as a consequence of what often amounts to a very slight genetic modification. In essence, the UPOV and TRIPs compatible systems will mean that seed sold in the developing world can be controlled after the point of sale. Patents and PBRs place strictures on the subsequent use of plant varieties by farmers and breeders. Perhaps the most famous and contentious example of these practices was found in Monsanto's insistence on contracts of use with American cotton farmers. The UPOV and TRIPs rules will further the commercial appropriation of the seed, and ensure that new seed must be bought each time the production cycle begins anew. Seed markets in the developing world are now being increasingly targeted by biotechnology companies, in the knowledge that these countries are facing pressure to provide IPR rights that will restrict on-site seed saving by farmers. As the section below argues the extension of IPRs to the South serves to reduce the commercial uncertainties involved in biotech companies extending their market there.

Monopoly interests and TRIPs

TRIPs was almost certainly a product of what was effectively an alliance between TNCs engaged in knowledge-intensive economic sectors and developed state members of the WTO. The United States in particular was responsible for heavy handed arm twisting of the main opponents of TRIPs – such as Brazil and India – via particularly aggressive trade legislation which targeted countries who were failing to protect the rights of US intellectual property title holders (Hobbelink, 1991; van de Wateringen, 1997). The USA's TRIPs negotiating team was itself heavily composed of corporate trade and intellectual property lawyers – who were used to effectively out-flank the poorly resourced trade missions of developing countries.

However, the most significant indication of corporate interests in the agreement can be found in the tripartite coalition of Japanese, European and United States industry groups.[1] These bodies combined to produce what was called the Industry Submission to the TRIPs negotiating group. This was the first-ever negotiating submission by a non-governmental grouping to a GATT Round, and its contents were closely reflected in both the US and EC submissions to the GATT secretariat in the same year (for summaries of these submissions see Beier and Schricker, 1989). Business interests had also been instrumental in choosing GATT (and not the more obvious World Intellectual Property Organisation, WIPO) as the multilateral framework by which to introduce a single international IPR regime. The availability of patents for living organisms in the developing world was one of the areas of protection that many influential TNCs particularly desired. Monsanto and other life sciences giants were ultimately successful in achieving this goal.

The TRIPs agreement is ultimately an expression of a convergence between the international system of patent rights for biotechnology products and processes, and the increasing commercial dominance of biotechnology by Northern-based transnational corporations. It is significant that the WTO contains no antitrust code by which monopolies resulting from IPRs can be countered. Monopolisation will doubtless be a feature of the TRIP's life-patenting laws, especially given the fact that the sectors in which biotechnology is used heavily (including: pharmaceuticals and healthcare, agrochemicals, chemicals, food processing and plant breeding) have already been characterised by a prolonged period of restructuring via mergers and acquisitions. This has largely occurred because the technology itself supplies a common scientific starting point to a number of productive activities across economic sectors. Significantly, the life-patenting system will be used to control the underlying knowledge base of the technology and, thereby, to freeze in the commercial dominance of these sectors by a system of exclusionary rights.

However, biotechnology has also introduced structural change in a number of economic sectors which also produces uncertainty for TNCs. The dynamic nature of the change in technology system introduced uncertainty into many established areas of production (as is the case in the pharmaceutical and commodities sectors), and an uncertainty which is exacerbated by the emergence of completely new ways of producing goods and entirely new products. TRIPs supplies a means of minimising uncertainty as knowledge-intensive products reach global markets, as companies can ensure that R & D costs can be recouped via legally sanctioned monopolies. In the instance of agricultural production, TRIPs will not only ensure that bio-engineered plants are not 'pirated' by Southern farmers, but will also slow down the rate of innovation by means of traditional breeding. This is another way of stating that patent holders can, thereby, control the third world markets that they are targeting with increasing vigour.

But TRIPs was to a large part justified because of the need to provide an incentive to small firms so that they would continue to innovate in the biotech sectors. Small biotechnology firms were in fact vaunted as 'competing shoulder to shoulder with large transnational corporations, thus guaranteeing a highly dynamic interaction responding to the real needs of the marketplace' (Hobbelink, 1991: 30).

This is very much a false picture of the current international structure of biotechnology. Larger corporations are more able to bear the long-term investments that are required for biotechnology R&D, and are better placed than small firms to bring biotechnology products to the markets. Smaller

firms still play a role as sources of marketable research, and corporate relations with them have been managed via equity investment or through
licensing arrangements. These links have particularly suited TNCs, as they
are more able to out-source research whilst commercially controlling the
knowledge which the small firms (and universities) generate. In many
instances the TNC strategy has been simply that of buying out smaller firms
when a promising product line has been developed (as was the case in
Monsanto's 1998 buyout of Delta and Pine Land Co. – the developers of the
so-called 'Terminator' seed). Life patents have certainly provided an incentive for TNCs to ratchet up their investment in the life sciences industries.

Recent analysis of the international structure for life sciences industries
has confirmed that the trend towards corporate concentration, first identified in the late 1980s (Kloppenburg, 1988; Hobbelink, 1991), has become
increasingly entrenched and global in scope. In 1996 an estimated 79 per
cent of all cross-border FDI was accounted for by mergers and acquisitions
(UNCTAD, 1997). The life sciences industries reflected a global trend, and
were characterised by a heady pace of strategic alliances and corporate
buyouts of both smaller and larger players. Mergers and acquisitions have
produced giant companies that now dominate production of the most basic
necessities of life. In 1997 concentration of ownership over key sectors
reached an unprecedented level. The top ten seed companies now account
for 40 per cent of the world's commercial seed market. Likewise, the top
ten pharmaceutical companies have a 36 per cent world market share.
Similar figures are apparent when one reviews the agrochemical sector
(RAFI, 1997: 1)

These elite groups of TNCs have also had a clear incentive to enter the
plant breeding industry, and developments in plant biotechnology have provided the scientific push and commercial pull for a pattern of diversification
amongst the leading TNCs into the seed sector. Monsanto, for example,
transformed itself from what was principally a chemical giant to a company
sitting on the top tier of the international plant breeding industry. This is
largely because developments in plant biotechnology are enabling linkages
between sectors that were previously discreet. Monsanto was able to
combine its massive interests in agrochemicals with rationally designed
receptors in crops. Plants can now be engineered to 'fit' with established pesticides and fertilisers. In general, new plants will also very likely be designed
to be suited to mechanised harvesting and the requirements of industrial
food processing, and plant breeding is also an increasingly important avenue
by which new drugs can be produced and delivered. A commercial and
scientific synergy is driving this flexibility, and this explains the voracious
manner in which TNCs from disparate sectors of production have bought out
seed companies.

This pattern has more recently extended to developing-country plant
breeding companies and commercial seed distributors. Events in Latin

America from 1997 to 1998 exemplify a global trend. Monsanto took a 30 per cent share of the Brazilian maize seed market and over 50 per cent of the same market in Argentina. Monsanto also acquired Cargill's international seed division in 1998, giving it seed testing and distribution networks in well over fifty countries. Other major players – such as Dow and Ciba-Geigy – have rushed headlong to purchase small seed companies in literally every continent (RAFI, 1998: 2).

The list of buyouts and mergers in the last decade is almost an endless one. The opening up of developing world seed markets to genetically engineered seed will blur distinctions between traditionally produced varieties of plants and their engineered counterparts. Patented genes will literally infect the world agri-biodiversity, since by the year 2000 'nearly all commercial seeds of all major crops will contain one or more bioengineered trait' (Wood and Fairley, 1998: 27). The availability of IPRs for both the engineered varieties and the genetic information will serve to restrict the rights of farmers to breed from this material, and be used as a mechanism to make them pay for its use. In terms of royalty payments resulting from the enforcement of TRIP's plant variety protection rules, the transfer of revenue from South to North will be all the more staggering given the relatively low incomes of the producers involved. Shiva conservatively estimates that plant patent holders will receive $7 billion per annum in royalties from Southern farmers, with revenues flowing to an ever-narrowing band of companies (Shiva, 1993: 127).

The plant-patenting system is already being abused in the USA and EC, wherein patent claims are not only of a broad scope (see below), but are also being used to exclude competition in agricultural production. One 1994 analysis of European Patent Office (EPO) patent awards indicates exactly the likely manner in which TRIPs will be used in the developing world. Prior to 1990, one-third of all applications to the body were made by Lubrizol, Monsanto and Ciba-Geigy alone. In addition, three-quarters of the total applications were either made by transnational corporations or their subsidiaries (Wells, 1994: 116).

However, it is not as if developing world farmers even wished to compete with biotech corporations. Nonetheless, this is exactly what TRIPs will force them to do on a playing field that is far from level. Sadly, patents will fail to stimulate plant breeding of a truly meaningful nature, since product diversity will never correlate to biodiversity. Life patents will affect diverse cultures and social relations of agricultural production, and displace traditional knowledge systems that have fostered biodiversity and fed the South. This will occur irrespective of the desires and democratic rights of those affected.

The development of biotechnology and broad biological patents

TRIPs' provisions do not set any limits on the scope of life patent awards. But because of the rights granted through patenting micro-organisms and micro-biological processes there are substantial reasons to anticipate that broad patent scope will negatively impact upon the future development of biotechnology. The ability of companies to gain broad patents in the field is already a cause of substantial concern in many biotech-related sectors (Roberts, 1994; Barton, 1995b). The practice is already prevalent under the patenting systems of the EC and USA. Roberts claims that these types of patents threaten to slow down or even stop the introduction of useful and important technology (Roberts, 1994: 371). This consequence of the availability of strong and broad-scope patents would also undermine a principal justification of the patent system – namely, that it increases the incentive for greater innovation. Indeed, broad patents not only give firms extensive monopoly powers over a given market, but block the entry of a range of competing products and processes. This has clear implications for the development of biotechnology (Ko, 1992: 778).

Examples of broad patenting in developed countries abound. In 1995 the USA's National Institutes of Health received exclusive rights over methods to remove cells from a patient, altering their genetic makeup and returning them to the body. This effectively encompasses (human) gene therapy at 'a very fundamental level'. Well-publicised criticism from the scientific community followed, both because of a potential block on research and the breadth of rights over a whole area of biotechnology (Coghlan, 1995: 4).

John Barton provides further examples of broad patents in the agri-biotech and pharmaceutical/medical sectors:

> The PTO [the US Patent and Trademarks Office] has accepted very broad claims in a number of important cases. Thus, the widely publicised Agrecetus cotton patent covered all genetically engineered cotton seed by claiming: cotton seed capable of germination into a cotton plant comprising of its genome a chimeric recombinant gene construction including a foreign gene ... Similarly the Harvard mouse is really the Harvard non-human onco-mammal patent. [the] claim covers: 'A transgenic non-human mammal all of whose germ cells and somatic cells contain a recombinant activated oncogene.' (Barton, 1995b: 606)

The scope of these two patents are breathtaking, and are by no means exhaustive examples of a wider trend in biotechnology (Ko, 1994; Roberts, 1994). The Agrecetus cotton patent was eventually successfully challenged, but would have allowed the company to exert control over the entire market for all transgenic cotton seeds. In a similar fashion the Harvard mouse patent gave rights over all non-human animals with bio-engineered cancer bearing genes. The patent therefore protects a research tool that is of fundamental importance to cancer research. Broad patents therefore cover extremely important products and research tools in biotechnology, and exclude others from markets and entire avenues of research. As a result, developments come to be the exclusive scientific and commercial property of the patenting firm.

Similar broad patents to those detailed above are very likely to be permitted under TRIPs as the agreement places no real strictures on the scope of patents that can be conferred on either products and processes. It is also probable that, since developed countries have been increasingly permissive in granting broad patents, then they will apply pressure on the WTO to maintain like standards when disputes arise on the issue. In addition, the companies who have gained these types of patents in the USA and EC will obviously expect their governments to demand like rights in foreign markets. The IPR jurisprudence of these countries might well be used as a form of legal precedent for judging the correct scope of biotech patents awarded under the different national IPR systems of WTO members.

Broad patenting scope threatens to stifle competition through the sectors on which biotechnology is having the greatest impact, and the practice would seem to be an obvious adjunct to the concentration of ownership within them and the commercial control of biotechnological knowledge. This results from the manner in which the patent system is already introducing commercial pressures on life sciences industries. In the first place the broad patenting increases the pressure upon firms to seek alliances with other patent holders. These arrangements come into being because many biotechnological techniques and processes need to be used in conjunction with each other, or because single (and patented) developments in biotechnology often have manifold applications across economic sectors. Patent rights over these often very basic techniques, research tools and processes lead to exchanges of rights (by cross-licensing arrangements) between the patent holding firms. Stopford and Strange (1991) characterised these types of arrangements as firm-firm alliances, and they are seen as a key feature of globalised production. A web of commercial ties serves to share out R&D costs and the uncertainties involved in the knowledge-intensive biotechnology sectors. This web of arrangements displays the evolution of a form of 'patent cartel' (or for Barton, below, a 'patent pool') which is exercising exclusionary rights over biotechnology, and control over its constituent knowledge systems and techniques. The international availability of patents will further serve to tighten the elite group of companies that are already dominating the life sciences industries:

> A firm with a broad patent is likely to license it to others with expertise in a variety of different markets that may be involved... There is a serious possibility that many of the broad and basic patents will be cross-licensed in what amounts to a global patent pool. This poses genuine antitrust and policy issues. Firms will feel no need to license it to those who do not have technology of interest to them. Thus, new entrants into the market may be excluded unless entrants are strong enough to develop important technology of their own and thus obtain access to the community. Moreover, public sector and developing nation entities may find it difficult to obtain access to technology to serve markets of little financial interest to the private sector. (Barton, 1995b: 614)

The entrenchment of these patent cartels over the commercial application and science of biotechnology is likely to be another consequence of the TRIPs agreement, especially as firm–firm alliances become globalised via co-operative ventures and cross-licensing. Globalisation of firm–firm cross licensing, after all, require a degree of harmonisation in the IPR systems of the firms' host countries – and these are exactly the conditions which the new WTO regime supplies. This would not only constitute an issue of global monopolies over the technology, but over the right to do science in the field, as patent cartels will as much exclude others from research as markets. Orsenigo and others have suggested that the development of biotechnology required the free international flow of scientific information, and the exchange of ideas and techniques (Orsenigo, 1989; OECD, 1988, 1989; Kenney, 1986). This was very much a result of the academic origins of the science and the fact that biotechnology has always moved forward as a result of a number of complex and interconnected innovations derived from a variety of disciplines (microbiology, biochemistry and cell biology to name a few). Freezing out public institutions and other firms from promising avenues of research and enabling research tools can only have serious consequences for the manner in which biotechnology is developed and ultimately deployed and used socially.

Broad patents threaten to slow down the process of innovation in biotechnology, as patents are best suited to specific product developments that are closer to the market, rather than to innovations and developments 'that might contribute to a variety of products' (Barton, 1995b: 614). For Ko, the extension of broad patents to biotechnology internationally will ultimately contradict the most basic liberal economic assumptions as to the role of patents in stimulating innovation:

> Courts, as well as commentators, seem to have lost sight of the ultimate purpose of patent awards – promoting technological progress. Traditional patent doctrine, developed as it was for traditional industries, does not necessarily achieve its purpose – the promotion of technological progress – when applied to industries, such as biotechnology, in which research is characterised as by unpredictability and randomness. (Ko, 1992: 804)

Conclusion

It is increasingly clear that the developmental trajectory of biotechnology is being governed by a concern with the 'bottom line' – that is the ability of TNCs to make profits out of the science in global markets. The patent system might in the future constitute the most important criterion by which firms decide on future research lines and product development. It is certain that firms are already gearing their R&D activities into areas of biotechnology where patents are readily available. In this respect, plant biotechnology firms have certainly no incentive to produce crops that are not genetically uniform. Cumulatively the patent route will mean that firms will forego work in areas in which biotechnology has a potential to benefit human health, the environment and productive activity in general.

When R&D is subject to corporate IPR lawyers seeking legal niches in which products can be developed and marketed, biotechnology will also lose much of its random character. Scientific freedom in the field is obviously another important issue. Hobbelink provides a particularly chilling example of the impact of patents on the development of a bio-pharmaceutical product:

> In 1987, top officials of the US biotechnology company Genetics Inst. Inc. gathered at their headquarters to settle an important issue: which version of a new clot-dissolving drug to invest in. With money to develop just one of four potential products, the company's scientists argued for the one that had had the most positive research results. They [the patent lawyers] pushed a drug that had not tested well, but would command the broadest patent. And they won hands down. (Hobbelink, 1991: 108)

In this example, legally driven product development relegated science to a poor second place. The loser here was the consumer of the inferior drug. But when looking at the future development of biotechnology in the post-TRIPs global economy, the ultimate losers will be those who not already taking a leading role in the biotech patenting race.

In terms of the broader knowledge structure and TRIPs, TNCs will clearly have an enhanced capacity to control biotechnology (and other new technologies) over global markets. The extension of IPRs to life forms will consolidate and globalise monopoly rights over biotechnology products and processes, whilst creating a structure of accumulation which is based on knowledge control. TRIPs will also extend these monopolies to areas of human life and production where intellectual property rules were previously either thought to be inappropriate or morally unacceptable. TRIPs introduces a completely new system of rights for seeds, plant varieties, animals and human genetic information, and, in the instance of the IPR laws of many developing countries, even pharmaceuticals. The costs of TRIPs are therefore likely to be the disproportionate burden of the South and the world's poorest communities.

The trend toward corporate control of biotechnology is occurring whilst the technology emerges as the common scientific starting point for a range of productive activities. This commonality has not only involved radical changes in which various sectors produce goods, but promoted cross-sectoral mergers and acquisitions between leading TNCs (and smaller firms). Life patenting plays an essential role in the short but dramatic history of the technology, and is crucial to any realistic appraisals of how biotechnology will be used socially and productively in forthcoming decades. Indeed, as biotechnology matures as a technology, there are good reasons to believe that the sectors of production that it will impact upon will expand, and like-wise the range of products. Corporate control of the knowledge base of biotechnology and the biological resources it depends upon will therefore provide these agents with a substantial degree of structural power over the knowledge and production structures (Russell, chapter 6, this volume).

Note

1 Composed of Keindarten (Japan), The United States Intellectual Property Com-mittee (whose members include Johnson and Johnson, Monsanto, Hewlett Packard and other knowledge intensive firms) and the UNICE which is the busi-ness and industry lobby organisation in the European Community.

6

Biotechnology and international political economy: actor-networks in the knowledge structure

Alan Russell

Introduction

Perhaps the key question for international political economy (IPE) in addressing technology is to ask what makes some technologies grow rapidly and diffuse through the global economy. Is there a technological determinism at play – an ongoing cascade of technologies derived from past innovations, in turn impacting on social and political relations. Or, is there a deliberate and directed selection of technologies, chosen for identified social and economic needs – a process of technological progress that must be understood in terms of human agency. There is little new in this dichotomy for IPE or international relations (IR) (Nau, 1974; Vogler, 1981). Nevertheless a charge can be made against IPE that it has rather neglected this problem (Talalay, Farrands and Tooze, 1997). Belatedly there have been moves to internalise technological change within IPE theory.

Two important contributions to understanding the role of technology within the concerns of IPE come from Robert Cox (1987) and Susan Strange (1994). Strange, especially, locates the inclusion of technology – as knowledge – in structural terms. Her general conceptualisation of a knowledge structure has great merit in the placing of biotechnology in IPE, and in some measure underpins the chapters by Williams (chapter 5) and Pownall (chapter 7). However, its structural orientation – or macro focus – cannot be expected to deal with the detail of individual technologies – which often have highly significant micro aspects – and, more fundamentally, an agency context. Biotechnology is an excellent example of technology in this context.

This chapter adopts the broad framework of Strange but then discusses two ways by which the knowledge structure, in relation to biotechnology, can be fleshed out. They represent two 'cuts' in this endeavour. Each cut is drawn from separate perspectives developed to internalise technology into socio-economic processes. The first cut utilises some of the established work of innovation economists – a group of scholars who have come to recognise the limitations of conventional economic analysis in its assumption that technology can be left external to economic theory and treated as a 'given' (Rosenberg, 1982, 1994; Dosi, 1984; Dosi *et al.*, 1988; Nelson and Winter, 1987; Foray and Freeman, 1993; Freeman, 1990; Archibugi and Michie, 1997). This literature offers a helpful lead for IPE in conceptualising technology from a 'micro' image of individual technology trajectories, or pathways, through to a level of pervasiveness where the whole global economy is affected. It is a self-contained literature, with its own debates, that offers IPE an off-the-shelf starting point.

It is, however, only a starting point for IPE because the innovation economists do not give sufficient attention to socio-political processes. The second cut offered as a means to flesh out the knowledge structure has its intellectual origins in the sociology of technology and is known as Actor-Network Theory (ANT) (Law, 1991). It also assists in bridging the micro/macro divide and offers interesting contrasts with the innovation economists. The two perspectives represent different cuts in an heuristic sense. They can both be applied to biotechnology and provide a basis for developing assessments of technological progression in their own terms. One cut sits nearer to traditional economics, although is critical of the latter's treatment of technology, while the second perspective offers more in socio-political terms, coming closer to broader IPE interests. However, ANT is less defined and less explicit in its conceptualisation.

Thus the work of the innovation economists will be one port of call in opening up a framework for the inclusion of technology in IPE (Russell, 1995) and ANT another. However, an explicit focus on 'knowledge' in the global political economy comes first.

Biotechnology and the global knowledge structure

Robert Cox and Susan Strange stand out in drawing explicit attention to the important part that technology plays in the concerns of IPE. As will be seen, Strange describes a 'knowledge' structure, within which sits technology (Strange, 1994). It is important, therefore, to stress that technology and knowledge are linked. As Skolnikoff observes: 'it is critical to understand that technology ... implies more than simply a piece of hardware; it implies the larger knowledge base that is specific to the creation of a piece of hardware and that made its production and application possible' (Skolnikoff, 1993: 14).

From a radical Gramscian perspective (Gill, 1993), Cox sees technology 'as being shaped by social forces at least as much as it shapes these forces' and is 'the means of solving the practical problems of societies. In this he is internalising technology. The selection of what problems are to be solved and the kinds of solutions acceptable are in turn determined by those who hold social power (Cox, 1987: 21). Social power is deployed with respect to social relations of production which have historically developed – to a point where there is now a co-existence of different forms in different states, or parts of the world. Social relations of production provide a focus for understanding processes at work in the global political economy, from the sub-state level, through the state/inter-state, to global levels. They entail what kinds of things are produced, how they are produced, the complementarity of roles played in the production process and the distribution of the rewards of production (Cox, 1987: 11–12). In effect, they involve the pattern, or configuration, of social groups engaged in the production process with 'power relations of production' observable in dominant-subordinate aspects of this pattern. Technology plays its part in all this, given its link with social power and production relations in general.

Susan Strange goes further than Cox and develops a more formal integration of technology, within what she describes as the *knowledge structure* of the global political economy – a structure which co-exists with and reinforces the three other primary structures of *finance, security* and *production* (1991, 1994). In effect they are seen as mutually supportive. Thus technology, as part of the knowledge structure, affects production, global finance and security. The knowledge structure is, however, more than simply technology. Important as technological knowledge is, it is usually deemed part of the 'instrumental' side of the knowledge structure, although the knowledge structure also extends to the realm of ideas and beliefs – including moral conclusions and principles derived from those beliefs (Strange, 1994: 122–5). Biotechnology engages the global political economy at high levels of instrumentality (Williams, chapter 5, this volume; Falkner, chapter 9, this volume). Indeed, since the early development of the new biotechnology, the instrumental promise of biotechnology was central – to the extent that it had a significant effect on altering industry–university links (Kenney, 1986; Ronit, 1997). Biotechnology was quickly seen by both industrialists and academia alike as heralding significant economic portent. However, this particular technology goes beyond instrumentality and has the capacity to raise profound questions of belief, ethics and principles as well (Dyer, chapter 3, this volume), making its potential effects on the knowledge structure extensive and its progression controversial.

Within the four primary structures the deployment of 'relative power' is perhaps unremarkable – as Strange does not depart from a common IR view of actors mobilising their resources to achieve their ends in competition with others. More interesting is her use of the term 'structural power'. This is significant in that it concerns the fundamental determination of the 'structures of the global political economy within which other states, their political institutions, their economic enterprises and (not least) their scientists and other professional people have to operate' (Strange, 1994: 24). Her approach is quite complementary to that of Cox who sees a structure existing from the level of social relations of production, through the state and state system level, to a global context. The deployment of relational power takes place within a given structure – in this case the knowledge structure. If relational power concerns the abilities of actors to get others to do what they otherwise would not – and encompasses the competitions and conflicts that characterise much in the way of the dynamic interactions of the global political economy – then structural power confers the ability to *shape* the structures within which the games involving relative power unfold. Specifically, structural power and authority within the knowledge structure 'are conferred on those occupying key decision-making positions' and 'on those who are acknowledged by society to be possessed of the "right", desirable knowledge and engaged in the acquisition of more of it, and on those entrusted with its storage, and on those controlling in any way the channels by which knowledge, or information, is communicated' (Strange, 1994: 121). Overall, the 'knowledge structure determines what knowledge is discovered, how it is stored, and who communicates it by what means to whom and on what terms' (Strange, 1994: 121).

Power in the knowledge structure is, however, diffused and not always expressed intentionally. Very often the effects of structural power in the knowledge structure are *non-intentional* and only recognisable by those on the receiving end of it (Guzzini, 1993: 461–2). This is especially so when the source of power is seen less in terms of lying with a single actor or territory but in terms of hubs in networks extending transnationally – which may centre on key territories like the USA or Western Europe. Thus what the USA does collectively, in terms of knowledge generation and control, affects all others. Such effects may create structural consequences or may involve simple competitiveness and relational consequences. Conceptualising a knowledge structure offers a means to include biotechnology within the concerns of IPE in an *endogenous* fashion.

The example of the Trade Related Intellectual Property Protection (TRIPS) provisions of the Uruguay Round give scope to consider *intentional* aspects of structural power in the knowledge structure and in the context of biotechnology as a technological paradigm (Williams, chapter 5, this volume; Russell, 1999a). In advance of any implementation of TRIPs, the United States supported its firms' interests by demanding bilateral arrangements, under Section 301 of the 1988 US Trade and Tariff Act, and amendments to the Foreign Relations Act, which included a provision requiring co-operative science and technology agreements to increase US intellectual property rights abroad. The USA targeted over thirty states, nearly all of whom conceded to US demands. For example, in late 1988, the US administration increased import tariffs on $165 million worth of goods from Thailand because of 'lax' enforcement of intellectual property rights. Also in 1988, Brazil faced punitive tariffs of some $39 million because it did not allow for patents on pharmaceuticals (Hobbelink, 1991:103). In both cases protection of biotechnology products and processes were significant objectives of the USA. In Thailand domestic groups responded to US pressure and attempted to prevent the parliament from amending the law. Fearing the potential trade losses the Thai government brought in legislation in 1992 which enabled patent protection for many aspects of biotechnology, bar the patenting of micro-organisms. Yet, the USA continued to apply pressure for further Thai reform to include the patenting of micro-organisms, plants and animals (Compeerapap, 1997: 13).

Thus in advance of multilateral proposals incorporated in the TRIPs part of the Uruguay Round, targeted states began revising their intellectual property protection laws (Wijk, 1992: 15). The USA was quite simply in a strong position of structural influence. The accumulative effect of this was finally manifest in the actual TRIPs multilateral agreement (Williams, chapter 5, this volume). This trade pressure strategy, encompassing biotechnology patents and plant protection issues, was also applied to other developing countries, such as Ecuador, India and Pakistan (Wateringen, 1997: 21).

Both trade-related intellectual property protection and biodiversity – which has achieved its high profile since the Convention on Biological Diversity – stand out as developments of direct importance for IPE (Williams, chapter 5, this volume; Vogler and McGraw, chapter 8, this volume). The control of resources, the process of technology transfer and trade are all affected. Biodiversity is also one part of a set of environmental issues surrounding biotechnology, which encompass both positive and negative aspects (OECD, 1998; Mantegazzini, 1986). High profile concerns extend to the restructuring of industries where, for example, we witness MNCs acquiring all previously independent seed companies, a spate of pharmaceutical mergers and new transnational inter-firm alliances. And, given its important regional strength, EU biotechnology policy has important consequences for Europe and relations beyond (Wheale and McNally, 1993, Pownall, chapter 7, this volume).

Much of the impact of biotechnology is transnational in character. The transnational trends of globalisation see biotechnology firms operating within complex sets of relationships described by Stopford and Strange as the 'new diplomacy'. Firm–government relations, firm–firm relations and the interactions of mixed groupings of firms and governments are all involved (Stopford and Strange, 1991; Pownall, chapter 7, this volume; Williams, chapter 5, this volume). The spread of societal and ethical concerns, sometimes erupting in vociferous debates, is again transnational in reach. In this respect it is worth mentioning concerns about safety – including laboratory practice, industrial scale-up, environmental release of genetically modified organisms and the controversy over genetically modified foods (Vogler and McGraw, chapter 8, this volume; Falkner, chapter 9, this volume). Moreover commercial drivers can take biotechnology in directions that raise ethical questions – such as the cloning of humans as an infertility alternative – which global society must resolve. If biotechnology is recognised as a new and important 'technological paradigm' (Dosi, 1984; Walsh and Galimberti,

1993; Orsenigo. 1989) with multiple trajectories, not confined to any one market, then the trajectories of development are leading to transnational changes in the social relations of production. Indeed, the whole process of technological diffusion and absorption in the global political economy involves transnational processes, including inter-firm networking (Farrands, 1997).

Whether we simply acknowledge biotechnology as a kcy technology for the twenty-first century (Arber and Brauchbar, 1998) or have specific interests in one or other of the component issues, there is strong reason to consider how biotechnology in the global political economy might be conceptually addressed.

Fleshing out the knowledge structure – a first cut

Innovation economists and Actor-Network theorists help us address the problem of this inclusion by breaking biotechnology and other technologies down into component elements of the knowledge structure. In a sense they offer us 'building blocks' through the concept of technological paradigm and other associated concepts. Innovation economists usefully identify *levels* of pervasiveness of a technology in the global economy stemming from individual trajectories or pathways, within technological paradigms, through technological systems to all pervasive techno-economic paradigms.

In relation to biotechnology we can start with specialist *technological trajectories* – encompassing agriculture, medical treatments, pharmaceuticals, energy, chemicals, environmental management, reproductive technologies. Each trajectory or pathway can be located in a broader *technological paradigm* (Dosi, 1984, 1988), sourced to modern developments in genetic engineering and other sophisticated means to manipulate DNA (Freeman, 1989; Walsh and Galimberti, 1993; Orsenigo, 1989). Because biotechnology is a generic technology (Dunning, 1993) these new technical breakthroughs can have extensive utility in a full range of industrial, agricultural and medical applications – modifying traditional methodologies and introducing new processes (Manning, chapter 2, this volume). A focus on technological paradigms includes generic developments, while technological trajectories represent more specialist applications or pathways. Akin to its scientific cousin (Kuhn, 1970) the technological paradigm facilitates stability and continuity – along component trajectories – until upheavals lead to a revolutionary new paradigm.

The next level of diffusion would be to describe biotechnology as heralding a new *technological system* which invokes an image of linkage of economic sectors or branches of the economy (Freeman and Perez, 1988). Technological systems are '*networks* of agents interacting in a specific economic/industrial area under a particular *institutional infrastructure* or set of infrastructures and involved in the *generation, diffusion, and utilization of technology'* (Carlsson and Jacobsson, 1993). The networks form around competence and knowledge flows and encompass organisational and managerial innovations – which, in the terminology of Strange, have a shaping influence on the knowledge structure. New products and technological improvements are extensive; costs of many services and products are reduced; social and political acceptability must be achieved; and pervasiveness throughout the economic system must be recognised (Freeman, 1989).

Biotechnology is heading towards becoming a technological system but has some distance to go (Freeman, 1989). In the longer term biotechnology might even become associated with clusters of technologies that are so pervasive that the term new *techno-economic paradigm* might apply – whereby 'changes in technology systems are so far reaching and pervasive in their effects that they have major influence on the behaviour of the entire economy' (Freeman and Perez, 1988: 47). This would put biotechnology in the league of oil technology or micro-electronics and has certainly not been achieved to date, nor likely to be in the foreseeable future. But it cannot be ruled out in the more distant future – if the twenty-first century lives up to the claims of some and becomes the biological century (Manning, chapter 2, this volume).

Such well-established concepts provide an initial basis to disaggregate the knowledge structure as far as technology is concerned (Russell, 1995). There is no denying the relevance of these concepts in helping us to understand how a pervasive generic technology like biotechnology can be internalised into market processes. The power of the concept of technological paradigm is that it is suggestive of the sheer innovativeness of certain new technologies – massively changing what went before – while the concept of technological trajectory draws attention to the multiple directions the new technology might take. However, it is important to stress that at all levels, from individual trajectories to techno-economic paradigms, the economic context must be viewed alongside a socio-political context – usually omitted in the innovation economics literature below the level of technological system – which by definition must contain far-reaching social changes. Yet, technological paradigms share some of the social or community elements of scientific paradigms (Kuhn, 1970; Metcalfe, 1997). Selection of trajectories and their success in biotechnology, with attendant changes in the social relations of production, are not just decided on an economic basis but by socio-political preferences and in relation to wider debates – that in the case of biotechnology were evident from the beginning because of profound questions of safety and ethics (Russell, 1999a). However, the technological community, compared with the community of scientists, is actually less cohesive, more differentiated and 'is more widely linked to other communities (e.g. manufacturers, government bodies) whose needs must be taken into account' as trajectories proceed (Balmer and Sharp, 1993: 474). Actor-Network Theory is more explicit in this regard.

Fleshing out the knowledge structure – a second cut

More radical insight on the transition from a micro focus on trajectories to macro effects such as pervasive technological paradigms, or even techno-economic systems, is to be found in Actor-Network Theory. This body of literature has origins in the sociology of technology (Law, 1991) but has impacted on the innovation economists (Callon, 1993), the study of human geography (Murdoch, 1997) and IPE. Busch and Juska (1997) have sought to address a biologically relevant case study – edible rapeseed production – through bringing ANT to bear on IPE analysis. They locate their analysis by pointing out that ANT can address a key weakness of traditional political economy by offering a means to bridge the macro/micro divide. Murdoch (1997) makes a similar observation about ANT and the discipline of Geography noting again the potential to bridge divides. These include the macro and micro levels as well as bridging other divides such as 'society' or 'nature', 'global' or 'local', 'state' or 'market'. This last divide takes us back to the heart of IPE (Strange, 1994; Schwartz, 1994).

The key issue, from an ANT perspective, is not one of scale. Where the previous discussion of disaggregating the knowledge structure was suggestive of differing scales from trajectory to techno-economic paradigm, ANT brings local and global together by focusing on connectivity (Whatmore and Thorne, 1997: 290; Michael, 1996: 68–9). Connectivity is also central to the notion of power in the knowledge structure, lying at hubs in networks which can extend transnationally. Whatmore and Thorne cogently identify the ANT stance:

> Where orthodox accounts of globalisation evoke images of an irresistible and unimpeded enclosure of the world by the relentless mass of the capitalist machine, ANT problematises global reach, conceiving of it as a laboured, uncertain, and above all, contested process of 'acting at a distance'. (Whatmore and Thorne, 1997: 290)

Michael takes up this discussion of scale and observes that ANT asks the question: 'How does a micro-actor become a macro-actor?' He argues this stands in marked contrast to alternate views that suggest macro-actors exist unproblematically. He notes that ANT avoids obscuring the fact that 'macro-entities are achievements and, in the process, affirm the power relations entailed by them'. Thus it becomes important to 'flatten' macro-actors 'into a series of micro-situations and thus to map out the multiple negotiations that necessarily contribute to the (re) production of the relations of power entailed in macro-actors'. Importantly, an actor which makes other actors dependent on itself, translates their wills into its own language, and renders these translations more durable, 'has grown in size' (Michael, 1996: 69), perhaps acquiring significant structural power along the way.

In biotechnology we have seen various actor-networks grow to prominence within the knowledge structure. These include: the network of scientists interested in developing the new techniques of genetic manipulation who, back in the mid 1970s, raised the whole issue of safety in the face of conjectured risks; and the enrolment to the growing network of specialist and professional institutions, both national and international (Russell, 1999b). Other networks grew which translated the impact of the new science differently: legislatures, activist groups, such as Friends of the Earth and the Sierra Club; and sections of the media. Into the melee would come companies (old and new), funding bodies, universities and growing sections of the general public. Gradually initial events and interactions which were very prominent in the United States and the United Kingdom spread internationally (Russell, 1988; Wright, 1994). Local debates attained national and international prominence and co-existed with the emerging international efforts to achieve some harmonisation of safety practice (Krimsky, 1982). One such example was the highly publicised Cambridge, Massachusetts, controversy of June 1976, centred around a proposal to establish a relatively high containment genetic manipulation research facility at Harvard University. The Mayor (Alfred Vellucci) was lobbied extensively by groups opposed to the facility. These included the Boston area Science for the People, Friends of the Earth, certain scientists including Nobel prize winner, Professor George Wald of Harvard University and his wife Professor Ruth Hubbard. Press coverage was extensive as the Cambridge City Hall became the centre of a fiery local debate in turn leading to over a hundred hours of hearings undertaken by the Cambridge Review Board, set up by the Council. Harvard University undertook two consecutive three month moratoria on building the facility before ultimately the Review Board concluded the facility would be safe (Grobstein, 1979).

In all, stories can be told here of the build-up of networks – some having the distinct character of possessing *sufficient leadership to act in their own right*. A very significant departure associated with ANT is its explicit assumption that the networks associated with technological activity are not just composed of humans but they are *heterogenous* networks, embracing both the human *and non-human* (Murdoch, 1997: 332). This is a major thrust of the ANT literature and of huge importance in the way we address technology. Moreover, individual actor-networks, including non-human actants, may in turn be networked together, forming larger-scale networks. Murdoch (1997), highlighting the limitations of too narrow an emphasis on social relations at the expense of social/material links, suggests the actor-network approach: 'gives us a much more mixed-up account of human/nonhuman relations as it tries to establish how the complex associations which comprise our world are predicted upon the compliance and adherence of material objects, technical artifacts, animals and humans' (Murdoch, 1997: 333). Thus not only is technology internalised into markets, as innovation economists argue, it is internalised into social relations as well.

Biotechnology certainly possesses a heterogenous character. The individual trajectories and the paradigm as a whole have been founded on new scientific techniques which produce artifacts (such as computer-driven machines to decode DNA or effectively create new strands); or which utilise biological chemicals with newly identified characteristics (such as the restriction enzymes that can splice DNA in a fashion that allows rejoining to any other sourced strand cut by the same enzyme); or interactions with non-human living organisms the subject of experimentation or production by engineered means (including Dolly the sheep or more mundanely a bacterial strain capable of producing human insulin).

As the technology of biotechnology has unfolded over the years these heterogenous associations of humans, artifacts and non-human life have shaped that technology, its paradigm and its trajectories. Consider Dolly the sheep, the first animal to be cloned from an adult mature cell. It was not produced to further the cause of cloning but to produce consistently reliable generations of animals, that could be made capable of producing drugs in their milk. Scientists interacted with the technology for this end. As the network associations grow from this development new interactions (or 'translations') of the technology will lead off in new directions (or trajectories) – perhaps in ways unforeseen by those who interacted with the components of life to produce Dolly. Another larger-scale example sees the social relations of production altering in a North–South context, as traditional agricultural practices are threatened with new genetically modified seeds (Falkner, chapter 9, this volume; Williams, chapter 5, this volume). The genetically modified seeds are, of course, technology artifacts that engage network relationships with farmers and companies. The operations of the firm alter and the practices of the farmer change as the latter are 'enrolled' or 'recruited' into the new networks (see below). Reward distribution from the whole process of interaction can change and there could even be effects in terms of changes extending to the knowledge structure as a whole – affected through the emerging technological system that is biotechnology.

ANT also encourages a temporal or historical perspective, a view shared by Cox and others in the field of IPE (Amin and Palan, 1996), whereby past action can incrementally influence present network identities and features (Murdoch, 1997: 329). Past layers of enduring heterogenous associations build up incrementally so that present action involves encounters with the labour of past or distant relationships. Structures are built up in this fashion at all times, accounting for non-human involvements, with legacies of past labour contained in material objects, with the link from present to future shaped by new objects derived from current heterogenous associations, all leading to a social fabric weaved around human and non-human actants (Latour, 1991: 103). Moreover, Latour stresses that contrary 'to the claims of those who want to hold either the state of technology or that of society constant, it is possible to consider a path of an innovation in which *all the actors* co-evolve' (Latour, 1991: 117). In other words, the technology artifacts evolve, the relationships with users evolve and the users themselves evolve.

At a minimum, debates today over the safety of releasing biological organisms into the environment, or the production of genetically manipulated foods, involve legacies of past developments and interactions. Had the earlier debates on laboratory safety – of high profile in the 1970s – not been resolved in a fashion to allow a moratorium on the use of the techniques to end and the technology to flourish, the contemporary debates would not now be happening. Specific technological breakthroughs can also be developed that in the longer term produce a possibility not recognised at the time of the development – such as lasers eventually being used to play popular music or movies, or developments in targeting cancer treatments to specific cells, identified by genetic marker, leading to potential ethnic weapons. But this is not the same as implying technological determinism. The debates today over genetic manipulated foods or genetically modified crops were not inevitable. The results of past labours must be translated anew in the present. Contemporary networks have coalesced around shared processes of translation, mobilised power and leadership, and present-day heterogenous networks, with their own unique characteristics at once linked to the past but also distinct, interacting also with present-day technological artifacts.

Thus the micro-macro divide is addressed by documenting the change of scale from micro to macro 'with the inclusion of greater and greater numbers of black boxes'. This encompasses 'the progressive re-opening, dispersion, and disbanding of actors passing from the macro level to the micro level' (Latour, 1991: 119). Black-boxing reflects the rendering durable of particular associations; they become 'unproblematic' (Michael, 1996: 54). In this some linkage with the literature of innovation economists is evident – as it posits increases of scale from trajectory, through paradigm to technological system – but the interpretation of progression differs. In this respect both innovation economists and actor-network theorists *offer insight into how technological knowledge and its artifacts may build up in the knowledge structure.* Technological trajectories and paradigms can be seen as networks – and specifically heterogenous networks – that, at all levels, bring in non-human actants. In biotechnology these non-human actants may be technology artifacts or non-human life forms, of all sorts. This takes us beyond Cox's focus on the essentially human social relations of production, that he identifies from the local to global levels.

Actor-Network Theory also requires us to deconstruct actors. Whether governments or corporations ANT suggests someone, a person, in each context of action, must act for the actor. Monsanto, in this respect would not be seen as a unified 'actor'. However, the person is not Monsanto either. The person is the person at that time, in a particular context, empowered to represent the network of (internal) relations that comprise Monsanto. In this respect Monsanto is an 'actor-network', both network and actor simultaneously (Callon, 1991: 142). Governments also disaggregate in the same fashion. As the scale increases, and it is apparent that complex networks attain the capacity to act, we should address the leadership dynamic within the network and the associated power to shape the consensus on translation, giving meaning to other actors. The network of firms, many with biotechnology interests, which lobbied the US and other Western governments in pushing for the TRIPs agreement, to the extent of comprising a large part of negotiating teams, is illustrative (Williams, chapter 5, this volume). Where a large-scale network is amorphous or lacking in leadership, structural effects may be non-intentional but significant nonetheless.

Moreover, social relations of production can involve changing associations that include technology actants. Technology, in this sense, becomes more than merely a component of the means of production, owned or controlled by some group or class. The concept of 'translation' is used by ANT to describe how actors in heterogenous networks view and understand each other (Latour, 1991; Callon, 1991). Thus the state of technology, its artifacts, both past and present, are given meaning in networks through the translation held by the different actors at any time. Decisive changes in translations represent upheavals and reconfigurations of networks, new mutual definitions, compromises, and may result from processes of negotiation involving human and non-human actants. Such changes can also accompany the development of a new technological paradigm, especially if a paradigm is recognised as having a socio-political dimension. Our preference to look for continuity in understanding relationships may detract from important points of transition. In the words of Latour, both 'economics and stable sociology arrive on the scene after the decisive moments in the battle' (1991: 120).

It is the points where actor-networks meet that represent the high points of the public record of events. Here we see the competitions between differing translations, different interests and the key points of bargaining. However, we must not lose sight of the process by which individual actor-networks develop. This process of 'enrolment' or 'recruitment' is very significant in following the history of a technology and its relationships (Callon, 1993; Busch and Juska, 1997: 695). Monsanto as a company developed over time, exploring technological niches and consolidating its operations around translations of technology and its artifacts. The emerging network in the UK which grew to challenge Monsanto and other companies in their development of genetically modified foods and plants have come into 'contact' at high profile points of disagreement and action. Nevertheless both actor-networks must coalesce around legitimised leadership to be recognised as actor-networks rather than simply networks. In this respect there is a story to be told of the anti-GMO lobby, its processes of enrolment and the consolidation of leadership. Both groups comprise network relations that hold different translations of the science, the artifacts themselves (GM foods and crops) and each other. For whatever reason Britain has become a focal point of concern over GM foods – a concern not shared to anything like the same extent in the United States. Yet, as we have seen, in an earlier decade, the US was the focal point for extensive debate – along the lines of the Cambridge debate – over the issue of laboratory safety.

Entering the knowledge structure: conclusions

The conceptualisation provided by both innovation economists and actor-network theorists help in the description of biotechnology emerging into the knowledge structure. Both provide a means of recognising continuities and discontinuities of progress, but with differences in emphasis. In this respect the advent of genetic manipulation techniques in the early 1970s heralded a new technological paradigm that was consolidated with the recognition of the commercial importance of the new techniques.

The new technological paradigm has been partly characterised by its research intensity, something which has encouraged the linkage between academic research and the concerns of firms. Clearly the *new* biotechnology has represented a huge leap in the possible compared with prior expertise in micro-biology, biochemistry and other specialisms. Conceptualising movement along trajectories resulting from this draws out differences between the innovation economists and ANT. For the former, trajectories represent continuities while changes in paradigms, the more generic level, represent the major discontinuities. For ANT the concept of trajectory over-emphasises continuity. Supporters prefer to disaggregate further to recognise upheavals in the individual trajectories, represented by changes in the interactions within associated actor-networks, including human–non-human interactions. Indeed ANT questions our tendency to focus on continuity and durability. Thus early progression saw considerable change and upheaval as those active in developing the new techniques came to enrol in a growing network concerned with the very safety of the laboratory work (Grobstein, 1979). This network saw continuous debate over the process of translation of the risks, reformulations of the assessment of risk and extensions of the network interactions to involve the media, environmental activist groups and legislatures (Russell, 1999b). There is a danger that the innovation economists miss the significance of these issues in the progress of trajectories and the technological paradigm as a whole.

Even while limited to a laboratory set of techniques to recombine DNA there was much publicity. This was largely due to the safety issue which quickly grew to international proportions, as scientists went public with their fears over conjectured risks. Open letters calling for safety assessments and a moratorium on work, until safety could be reviewed, drew public attention to the new technology (Grobstein, 1979). With the media reacting to the concern voiced by the scientists, legislatures soon followed. In the United States the debate was particularly fierce. However, set against this was a general consensus amongst the scientists that biotechnology promised a huge volume of long-term rewards – and a good few in the short term.

The balancing act between risk and reward took on the characteristics of a socio-political process. The various networks which developed – in ANT terms – took different perspectives on the technology, itself advancing rapidly. By the end of the 1970s some 30 countries had begun work on a safety regime for laboratory research often based on the US Guidelines or the British approach. The former produced written physical and biological containment guidelines for experiment types while the latter involved a national committee, before which were put all proposed experiments involving genetic manipulation for a case law recommendation on appropriate containment (Russell, 1988). Moreover, transnational networking developed, initially between scientists, through their conferences, professional organisations and associations. The transnational networks extended gradually to include influential organisations like the European Molecular Biology Organisation, the European Science Foundation, the International Council of Scientific Unions, the International Association of Microbial Societies, and the World Health Organisation (Russell, 1988).

In the meantime, industry began its adoption of the new techniques as the technological paradigm grew and trajectories consolidated (Elkington, 1985; Daly, 1985; Orsenigo, 1989), creating their own networks and leadership on some issues. Local changes and translations operated within pharmaceutical companies as they deployed new expertise and closed down old practices. New arrangements between universities and biotechnology firms have altered conceptions of basic research as companies provide funding in return for exclusive rights to exploit any commercially viable discoveries or innovations. The transition towards greater reliance on advanced skilled labour, in what is a knowledge-based industrial activity, is associated with such changes (Hayward, 1997). Movements and localised shortages of scientifically skilled labour also suggest global issues of relations of production (Alic, 1993).

For example, many Japanese travelled to the US in the 1980s to acquire skills in the new biotechnology through doctoral programmes. More broadly, in Cox's neo-Gramscian terms we can describe a transnational capitalist business class deployed in biotechnology – leading to transitions in North–South agricultural production relations, purchases of specialised skills through establishing contractual relations between established large firms and new biotechnology start-ups. Both Japanese and European firms have been very active in such network links alongside large US firms (Pownall, this volume; Wiandt and Amin, 1994; Prevezer and Toker, 1996). Networks also bring firms and the public sector together. In recent years, the Senior

Advisory Group on Biotechnology emerged as an industry interest group that has been active in lobbying the European Commission on the direction of EU biotechnology policy, including safety regulations (Wheale and McNally, 1993).

Growth of the industry and its diffusion can be viewed in terms of the innovation economists or more along the lines of Actor-Network Theory. Either way IPE benefits in fleshing out the importance of biotechnology in the knowledge structure. From localised beginnings, the techniques were soon widely adopted and combined with existing and other new techniques, such as cell-hybridisation, in a technology which promised both fundamental breakthroughs in our knowledge of DNA and genetic processes, as well as commercial rewards across a number of industrial and agricultural activities.

The United States took an early lead. Favourable tax breaks for technology start-ups and the availability of venture capital has seen a huge explosion of new biotechnology firms, with today about a thousand in California alone (*The Economist*, 1995). These start-ups usually began as research firms with a long-term objective of transforming into fully integrated manufacturing and marketing firms (Daly, 1985: 17). They included firms that acquired early significance, including Genentech, Biogen, Cetus and Genex. Moreover, the United States has led the world in its national science base which has seen biotechnology attract high levels of public funding. In terms of structural power, other parts of the world have seen developments in the new technology, shaped in many respects by this US lead. The biotechnology industries of both Japan and Europe depend considerably on transnational linkages with US firms, which are in turn plugged into the thriving US science base. Indeed, for biotechnology most of the significant transnational network relations ultimately pass through the United States – the main source of both intentional and non-intentional influence over the global knowledge structure.

7

An international political economy perspective of the biotechnology industry: developing regional strengths

Ian Pownall

Introduction

The biotechnology sector offers the potential for European industry to effectively compete and regain lost technological advantages against Japan and the USA. This chapter compares the European responses to increasing global competition in the biotechnology sector against a changing industrial structure and political rationales. This is undertaken using an international political economy (IPE) framework initially outlined by Stopford and Strange (1991) and later developed by Mason (1994) and Lawton (1997). From a review of industrial structure, competitive needs of firms and policy measures developed by the European Union (EU), European nation-states and rival regions of the world, it is argued that the biotechnology sector is undergoing a transitional period and is shaped by two complementary themes:

- that nation-states and regional blocs, through firm interconnections, are seeking to develop and institutionalise existing political and cultural ties to maximise their international competitive position
- that recognised existing weaknesses in the biotechnology profile of a nation-state are being positively and aggressively addressed without (as of yet) overt concern to that competitive position.

There are four major sections to the chapter. The first section introduces the framework used to analyse the development of the biotechnology industry and measures implemented by the different nation-states and regional blocs. The second section describes the nature of the biotechnology industrial structure, whilst the third section merges policy measures with industrial structure dynamics to highlight examples of increased concentrations of structural and relational influence in the biotechnology sector. The concluding section attempts to fit these observations into the IPE framework, whilst suggesting future directions for European biotechnology policy-makers.

The international political economy framework

The analysis of globalisation issues inherently involves a consideration of systems and systemic effects (Humbert, 1994). Yet such approaches, especially when considering the business and management arena, rarely acknowledge the presence or location of nodes of structural or relational power within the industry. More importantly, this can fail to acknowledge policy constructed and location advantages within the firm's business environment. Figure 7.1 displays Lawton's (1997) adaptation of Stopford and Strange's (1991) new diplomacy framework, which views the formation of policy measures as being systemically driven between the firm and the state. In Lawton's adaptation for EU policy measures, the EU Commission through its *Directorate-Generales* (DG), becomes a major policy-making actor to consider and hence the original triangular diplomacy framework is replaced by a pentagon structure. However, as Mason (1994) illustrates in his analysis of the development of the EC–Japan Automobile Accord (1991), it does not necessarily follow that all the outlined systems provide a significant contribution to the formation of policy and the shaping of industry.

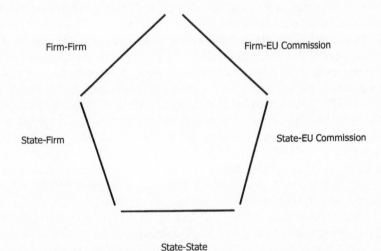

Figure 7.1 The Lawton (1997) Pentagonal Diplomacy Framework (EU focused)
Source: Lawton, 1997.

The model readily identifies the major communication and policy-making possibilities that arise at a firm–governance interface from a single-sector focus. With multiple sectors and policy streams potentially involved in the biotechnology industry, the location of idiosyncratic competencies amongst different regions in the world is not fixed and is more open to effective government policy direction.

Sklair (1998), for example, notes that any global system is supported by at least four pillars of firm–governance activity. The first and arguably most important of which is the extent and direction of Foreign Direct Investment (FDI) – helping to indicate the structural and relational characteristics of the biotechology sector. In addition, different actor capacities and the convergence between technological, competitive and cohesive issues have thrown greater emphasis upon welfare rationales such as employment, small firm dynamics and the function of the bio-entrepreneur within industry (Adam, 1997; DG XII 1997). As a result, this can be expected to emphasise different sides of the pentagon under different conditions.

Moreover, within the EU, the need for multiple nation-state support is reflected strongly in the continuing nature of its technology policy. Largely single-issue technology concerns have tended to result under Article 235 of the EEC Treaty (acting as a surrogate for an EU industrial policy (Farrell, 1996). It is only with the Maästricht and Amsterdam Treaties that the EU institutions have gained formal and codified competencies in industrial policy (Article 130 Title XII) (Nicolaides, 1992; Kipping, 1997; DG XII 1998). Nevertheless, the EU industrial arena still operates with a more functional, incremental and co-ordinating rationale (Holland, 1993; Adam, 1997). Member-state policies, therefore, remain important in shaping national industrial competitiveness.

This continuing role of the nation-state embraces the theme that the home base of the firm remains the core source of its competitive advantage. Firms retain their key assets and resources in their home country rather than a host nation-state (Ring, Lenway and Govekar, 1990; Wyatt-Walter, 1995; NSF, 1996; Dunning and Cantwell, 1991). While this is changing for the biotechnology sector (Dalton and Serapio, 1995), the competitive advantage of firms is generally maintained within 'nationally' controlled parameters. Peterson (1994) and Cerny (1996), for example, develop this argument with the concept of the 'competitive state'. Moreover, US international firms, in the high technology industries, continue to perform 90 per cent of their basic research within the US (Dalton and Serapio, 1995).

Thus regions, nation states and regional groupings of nation-states improve the prospect of developing and shaping structural and relational power attributes in global technological markets, by facilitating advantageous competitive conditions and factors for enterprises within their borders. Basic knowledge, as Strange (1988) and Metcalfe (1998) argue, is the key both to an understanding of the potential for any policy to promote and sustain competitiveness – in this case for biotechnology firms – and its integration into the process of policy development (Campanella, 1995). There may, therefore, be a better fit between the demands of the biotechnology industry and policy initiatives for certain societies than others. By focusing upon policy priorities (public and private), the scope of building regional competitiveness in the biotechnology sector can be gauged. This in turn provides a method of identifying sources of structural and relational strength in the biotechnology sector.

This fit arguably begins with historically accumulated biotechnology knowledge, for which the European states possessed an early advantage. However it may also be driven by societal concerns, such as an ageing population, particular diseases and regional and geographical needs, that, by accident rather than design, engender certain knowledge bases and skills to a given territory. Such a view of the biotechnology industry is arguably more relevant than adopting a purely competitive perspective for firm operation and formation (Whitley 1994; Nugent and O'Donnell, 1994; Burton 1999). This is largely because the source and fixing of power in policy dimensions – that facilitate total factor productivity (TFP) for the biotechnology sector – tend to be overlooked in orthodox business environment analyses. For example, both Metcalfe (1998) and a recent Asia Pacific Economic Co-operation (APEC) (1998) report, identify systemic integration of innovation factors and regionally specific societal concerns as having important contributing roles in competitive modern business.

The biotechnology sector's industrial structure

In a sector of the economy that is reliant upon new knowledge, basic research is a key competitive element. Associated with this are long industrial lead times and investment requirements. As a result, commercial success from biotechnology companies is difficult to predict with the sector prone to spectacular failures (Welles, 1995). In 1994, for example, US drug company Synergen lost $715 million of its market value in a single day when its Antril drug failed in clinical trials. Indicative of recent UK difficulties in the sector, shares of the small biotech diagnostics firm Shield Diagnostics have ranged from a low of 103 pence to a high of 919 pence (Newsedge, 1999a, 1999b). Increased merger activity between the large firms is also apparent with deals such as Monsanto's purchase of Dekald Genetics for $2.3 billion, which was swiftly followed by Du Pont's buy-out of Pioneer Hi-Bred International Inc. for $7.7 billion (Kupper, 1999; Holland, 1999; Morgan, 1999; Newsedge, 1999c). Such activities also serve to squeeze out the medium-sized firm from the sector (Powell and Pearson, 1995).

This, however, has not dampened enthusiasm for policy-makers to view the sector as having significant potential for growth and an ability to enhance regional technological competitiveness (FCCSET, 1992; DR Report, 1996; DTI, 1999; Kyriakou and Gilson, 1998; LaFee, 1999; DR Report, 1997; Berliner, 1999; Newsedge, 1999d). Overall, the sector is structured by an acceptance of long lead times, decreasing investor funds (and confidence), increased rationalisation and problems of intellectual property and healthcare policies.

Despite these obstacles, European biotechnology firms and policy-makers have reason for cautious optimism. Compared with information technology, for example, the dispersion of regional biotechnological strengths in Europe

is considerably wider and historically embedded. Europe has strong traditional biotechnological activities in fermentation, enzyme production,
agriculture, food processing and pharmaceuticals which have given the
large European transnationals a significant competitive head-start and a
locally accessible source of skills and knowledge. In particular, the UK, the
Netherlands, France and Germany dominate the biotechnology sector in the
EU. However, in a sector where new-to-the-world knowledge is perceived to
be a primary source of competitiveness, a static, or introverted, regional
market perspective means advantages can be quickly lost. The 1997 report
on '*Science and Technology Indicators*' from the National Institute of Science
and Technology Policy (NISTEP) in Japan, indicates, for example, that the
drugs and medicines industry maintains the highest R&D expenditure per
sales ratio of any industry sector, with 8 per cent minimum reinvestment.

Importantly, in the biotechnology sector, major constraints (and in some
cases advantages) result from the regulatory frameworks associated with
drugs and medicines, which have been established for public health reasons.
Notably this includes the role of the state as a significant consumer (Powell
and Pearson, 1995; Institute of Prospective Technology Studies, 1997).
Welles (1995) describes the constraints in this lucrative sector as:

- clinical failure;
- clinical success but market failure;
- delays in delivering products/research.

Whilst market failure is a risk of all businesses, the three-tier clinical
market demand system produces market imperfections. With the doctor
prescribing the drug, the patient taking it and the insurance company meeting the cost, neither the final demand nor the payment is controlled by the
consumer. This mismatch between market supply and demand forces is
further exacerbated by the nature of drug manufacture and clinical trials.
Pre-clinical laboratory tests must support an application for full clinical trials which have three phases. Each phase of the trials is subject to rigorous
regulatory control. Phase I trials involve assessing safety aspects with
healthy humans. Phase II involves administering the drug to a controlled
limited patient population, whilst phase III expands this limited trial to a
broader population in different geographic sites. Whilst phase I and phase II
trials can often be supported internally by the company, phase III trials
require substantial territorial/global support and manufacturing capabilities.
Only a limited number of firms are able to supply such resources. Contract
manufacturing is therefore a key competitive factor in drug development
(Werner, 1998; IPTS, 1997). Indeed, these essential supporting activities
include corollary research programmes, collaborative activities, information
management and quality assurance (Werner, 1998). Most small biotechnology firms can only carry as few as a handful of separate products lines
(Werner, 1998). They are limited in their potential to diversify product

ranges, given capital development costs needed for state-of-the-art facilities, ethical concerns over testing (and viable alternatives), pressures upon salaries and the costs of purification of a biotechnology product (Powell and Pearson, 1995).

Furthermore, for an EU-oriented biotechnology firm, coping with fifteen different nations, their public healthcare policies and regulatory requirements, adds to the complexity and market difficulty of generating new products. A supportive institutional environment that readily works with the firm on clinical testing issues arguably endows that firm with a distinct competitive advantage (Wolf, 1999; Thumm, 1999). For example, Singapore is recognised as being able to supply all the necessary resources for biotechnology clinical trials, which in conjunction with an aggressive funding regime ($2 billion devoted to biotechnological R&D in the current National Technology Plan) enables indigenous firms to rapidly gain market share through resource exchanges with foreign firms (BioAsia, Monitor 1998). In contrast, the Federal Drug Authority (FDA) in the US, has experienced budget increases that have largely been eaten up by inflation (Wechsler, 1998). Nevertheless, it remains a significant structural gatekeeper for access to the large US market (IPTS, 1997).

An additional consideration in the sector, is the changing nature of business rationales. Welles (1995) argues that biotechnology firms are a mix of three types of enterprise. These three enterprise types necessitate different policy measures (Fritsch, 1995). Implicitly therefore, the structure of the industry demands a mixed-strategy approach to policy development and measures.

Biotechnology firm types

The first generation of biotechnology firm (type one) is an integrated research and development (R&D) venture, concerned with in-house research, funding and marketing. The second generation biotechnology firm (type two) operates as a drug discovery company, more concerned with bioprospecting to find a new competitive advantage than leaving all research to an in-house team (Kupper, 1999). Finally, the third generation of biotechnology firm (type three) is argued to be the drug development company.

This last type of firm, is not concerned with direct in-house nor capital-intensive research, but directs its attention to the already abundant knowledge that has been made available through existing research and which has not yet been commercially developed. Such firms tend to maintain extensive links with University R&D staff, whilst operating as a virtual entity. This has less capital requirements, facilitates a greater product range and hence has less associated market risk for the firm. The research and skill emphasis changes from cutting edge new-to-the-world knowledge and speed of research to methodical examination and insight. This demands highly skilled individuals with extensive experience and an ability to recognise potential. As such, these firms emerge in developed countries (DC) with significant competitive strengths.

Type three firms, however, are not just small firms. Powell and Pearson (1995) give the example of Hoffman la Roche and Co. which introduced and marketed an antibiotic, Rocephin. This had already been developed by other firms but not previously marketed or recognised as valuable at the time. Arguably, Williams' (1998b) research boutique theme of small biotech firms, remains valid, but the small firms themselves are changing market and research focus as the industry matures.

From a holistic policy perspective, the type one firm is heavily dependent upon three dimensions of the support environment

- availability of capital;
- availability of skills and knowledge (including databases, journals, access to talented individuals and so forth);
- clinical and healthcare requirements (for biopharmaceutical ventures).

This places demands upon policy measures, in terms of financial and knowledge support, whilst operating a 'friendly' clinical environment. The UK, Canada and Australia, for example, are currently investigating methods of economically evaluating new drugs for their welfare services amongst broader healthcare reforms (IPTS, 1997), whilst the German clinical trials environment is argued to be driving investment out of the country and to the US (Slater, 1996a, 1998). In the US, the Prescription Drug User Free Act (PDUFA) is using revenues to finance drug evaluations and increase the rate and acceptance of new drugs (Wechsler, 1998).

On average, new medicines take around seven to twelve years from concept to market sales and cost around $300 million to develop (Thumm, 1999; Werner, 1998). In Japan, the regulatory environment for pharmaceutical products, with its intensive internal competition (through approximately fifteen hundred domestic firms), mandates that the prices for drugs must be reduced every two years, creating an incentive for new product development by necessarily ensuring short product lifetimes (BioAsia Monitor, 1998). Indeed, very little export income is generated by these firms, which serve primarily the second largest domestic market in the world (see table 7.1). Submission of new drug applications (NDA) by foreign firms have been greater than domestic firms since 1990 in Japan, as EU and US companies sell directly to the market (Kawamura, 1998).

As Thumm (1999) further notes, the regulatory framework and a supportive patent environment provide sources of significant competitive advantage for a biotechnology firm, protecting their intellectual property and encouraging and attracting investment. As an indicator of the potential value of this fact, and supported by the recent difficulties over the European Medicine Evaluation Agency (EMEA) (discussed later), an analysis by the

Table 7.1 Japan's technology trade (1995 – 100m Yen)

Sector	Exports	Imports
Motor vehicles	1,591	75
Iron and steel manufacturing	169	42
Drugs and medicines	367	367
Communications and electronics	1,528	1,734
Other industries	1,699	1,533
All industries	5,621	3,917

Source: NISTEP, 1997.

IPTS of the biopharmaceutical industry illustrated that no EU firms were included in the top fifteen of global firms patenting biotechnological research.

In terms of funding, the UK medical and biotechnology start-ups accounted for 21 per cent of 1990–94 Corporate Venture Capital (CVC), second only to information technology (IT)-based start-ups (McNally, 1995). However, a significant provision of CVC in the UK to these firms came from the US and Japan as UK banks and the British Venture Capital (VC) industry are more geared towards a role of replacement equity. There is a preference to buy out existing shareholders, rather than provide new equity (especially start-up and seed capital). Moreover, public monies tend to be sectorally, project and time structured, which is not often in sync with the growth objectives of a technological business (Mason and Harrison, 1996; Pownall, 1998; Durrani, 1996).

To counter national shortfalls, informal VC has emerged through Business Angels, as a dominant source of funding for type-one firms in the US and less so in the UK. However, funding is extremely scarce in developing countries, making it very hard to launch and sustain new biotechnology firms (NBF). Andrews (1998) generally notes that, in the EU, the investor situation is gradually changing in favour of technological interests, somewhat against the global trend. Yet there is a considerable way to go to catch up with US dominance in commercialising biotechnology research (IPTS, 1997). Indeed, recent reports suggest that the increase in available private funding is still largely geared towards IT ventures. For example, in 1994, only 2 per cent of VC was invested in biotechnology enterprises in the EU, compared with 25 per cent in the US (IPTS, 1997; CEC, 1997a).

For type two firms, more concerned with gaining a competitive advantage by casting a wide knowledge net rather than the narrow net with type one firms, capital requirements, whilst high, would arguably be more dependent upon prospecting success. This has implications for the intellectual property regime of the host and home region/state, the interpretation of a common natural heritage and the preservation of biodiversity (DaSilva and Taylor, 1998).

Existing national and regional patent rights are difficult to enforce and validate (European Federation of Biotechnology, 1996; Thumm, 1999). Negotiating international agreements, such as the Biodiversity Convention, with its difficult sovereignty issues over the ownership of genetic resources, the complex negotiations behind TRIPs and the slow development of the EU Directive on Protection of Biotechnological Inventions (initially proposed in 1988 but only accepted in 1998), illustrate this (Williams, chapter 5, this volume), for example. There have also been alleged cases of industrial espionage concerning San Francisco biotechnology firms, which were subsequently investigated by the FBI (Barnum, 1993). The alleged culprits were suggested as being Pacific Rim companies (although this was not proven). This highlights the continuing importance of effective knowledge appropriation and protection. The European Patent Convention (EPC) is still far from providing a harmonised legal framework which is further complicated when issues of patenting genetically modified organisms (GMO) are considered (Thumm, 1999). These are further key factors that shape regional competitiveness.

For a type three firm the same risks are run for bringing a product to the market. Within the EU, the creation of EMEA, ostensibly aimed at easing and facilitating the licensing of new medicines in Europe, has only largely added an additional governance layer to an already complex clinical and intellectual environment. This partly reflects the European pharmaceutical industry's lack of co-operation in implementing an industry-sourced solution to the problem (Slater, 1996b). It is also symptomatic of an acquiescence by European firms, and member states themselves, that this global industry is more competently handled by a global actor such as the EU, following in the footsteps of the evolution of the EU's IT competencies in the 1980s (Sandholtz, 1992; Molina, 1995).

Both type one and type two firms face significant purification problems with new products where these downstream separation costs can be the most expensive stage of new product development (NPD). It is, therefore, no surprise that bio-processing is a key element in Triad biotechnological support policies (National Science and Technology Council (NSTC, 1999)). For smaller firms, lacking in-house manufacturing capabilities, the biggest source of competitive advantage is arguably to outsource to a partner capable of undertaking the complete manufacturing process (Werner, 1998). Equally, the development of supporting bio-informatics programmes is an important prerequisite of a developing biotechnology industry that is heavily dependent upon rapid analysis of the volumes of data generated (Davey, 1998). Gaining additional insight and knowledge in these areas would constitute a significant source of competitive advantage.

The type three biotechnology firm has significantly different requirements and needs from its business environment. With reduced in-house capital needs and fewer problems of appropriating distant knowledge (through bioprospecting and out sourced research activities), their main policy support reliance is upon infrastructural routes for knowledge dissemination, communication and co-operation. Delays, arising from clinically rejected or market-driven causes, are less of a problem for a company that pursues many different product lines. Type one or two firms, with substantial resources, can obviate some of their difficulties by using type three firms, or start-up ventures, as technology transfer conduits (Powell and Pearson, 1995; BioAsia Monitor, 1998). One manifestation of this, the Strategic Alliance, has, aside from a dip during the recessionary period (1990–92), continually increased in the biotechnology sector since 1977 and sees no sign of diminishing (Sulej, 1998) (see figure 7.2).

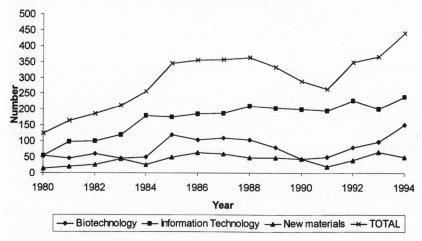

Figure 7.2 Strategic alliance formations between Triad technological firms (1980–94)
Source: Lawton, 1997.

Thus all three types of firm, which range from the large established biopharmaceutical style venture, with in-house R&D, to the small drug-developing company, necessitate a graduated support environment. Type one biotechnology firms maintain the highest dependency upon capital and clinical success of new products, whilst having a decreasing dependency upon knowledge dissemination in the sector and a decreasing number of mature products. However, type three biotechnology firms have an increased dependency upon knowledge dissemination in the sector with greater levels of mature market products. They also rely less upon capital and the clinical success of new products. Arguably there would be no specific relevance to size of firm, nor sector of biotechnology activity, except in the sense of the number of different research product lines that can be pursued simultaneously. In the agri-seed business, for example, the market is dominated by a handful of large firms whilst R&D is mainly led by the small firm (IPTS, 1997).

The international political economy of biotechnology

Daza's (1998) review of the United Nations University Biotechnology for Latin America and the Caribbean (UNU- BIOLAC) initiative is an example of the mix of firm–governance interfaces and policy measures in the biotechnology sector. This review stressed the importance of gender and ideology as issues shaping activity in this programme. DaSilva and Taylor (1998) similarly outline a series of socio-political factors that shape the effectiveness of biotechnology development programmes in the Caribbean and other developing countries. Furthermore, reports, from the National Science Foundation (NSF, 1998a) and the Japanese Ministry of International Trade and Industry (MITI, 1998) on current Japanese R&D policy, identified that key cultural changes are needed in perceptions of work and values maintained by Japanese society, in order to sustain its present competitive positioning in biotechnology (see also Dorabjee *et al.* 1998). This is largely based around a transition to a more meritocratic society, less ready to identify age with superiority or wisdom. On a more practical note, the North Carolina Bioscience Industry Forum (1998), home to the worlds largest enzyme facility and the Canadian Biotechnology Human Resources Council (BHRC) (1998), are active in promoting education initiatives at high schools (pre-tertiary level) (Magee, 1999).

The mix of societal policy instruments, addressed at culturally specific concerns, suggests that biotechnology policy has twin objectives. This first objective is to institutionalise an internationally competitive position whilst, secondly, regions themselves may be implicitly addressing biotechnological weaknesses, because of non-economic concerns. Nevertheless, total factor productivity (TFP) competencies, in those areas, are being accumulated by individuals, firms and governments. The EU's Pacific Regional Agricultural Programme (PRAP) is an example of a policy structure which serves both local competence development as well as strategic objectives of the EU.

The PRAP programme, operational since 1990, is comprised of a number of projects and is based in the eight African-Caribbean-Pacific (ACP) countries of the *Lomé* convention. The Caribbean countries had, through UNESCO's Microbial Resource Centres network (MIRCEN), the Caribbean Development Bank, the Latin American Energy Organisation (OLADE), the South Pacific Commission's (SPC) plant protection service and, with the development of a regional strategy in 1988 (*Perspectives for Biotechnology in the Caribbean*), implemented initiatives that used recognised traditional strengths in microbial biotechnological practices (DaSilva and Taylor, 1998). These included:

- bio-digester designs;
- fermentation;
- bio-remediation;
- tissue culture.

These are key issues being developed by the PRAP programme. Phase I sought to increase the regional facilities and infrastructural capacities, whilst the current focus of phase II (with a budget of nine million ECU) is upon developing national capabilities, ostensibly via training and education. Tissue culturing has been identified as a prime focus for supporting biotechnology initiatives in Japan, India, the Philippines and Australia (*BioAsia Monitor*, 1998, Science and Technology Agency (STA), 1998a). Maintaining access to such potentially crucial knowledge is arguably a major consequence of this initiative for EU firms. Furthermore, Vietnam, the Philippines and China are engaged in funding and despatching scientists to projects which were initially operated by PRAP solely for regional institutions.

Daza (1998) also notes that apart from EU sponsoring interests in programmes like PRAP, with structural intentions, a source of relational power has emerged in such areas through possession of considerable biodiversity. Yet developing countries are just gaining the skills to be able to utilise and protect this natural regional heritage. He describes this as a wider regional interest, or a South–South development. Whilst the UN originally created the BIOLAC programme for biotechnology development in Latin America (narrowly geographically defined), both Canada and Peru have become involved in the initiative (and contributed funding to it), with an intent on gaining knowledge from the programme's three main objectives:

- diagnostics;
- vaccines;
- plant genetics and micro-organisms of industrial interest.

Nevertheless, Daza's review of the origins of participants and the results of the programme supports a primarily regional perspective, with South–South networks of collaboration and information exchange dominating. Links with the developed countries, such as the US, are maintained but not intensively.

The reciprocity requirement of USAID programmes in biotechnology activities (NSTC, 1999), also shapes the form of developing country markets. Similarly, EU programmes to ACP and Central East European (CEE) states as well as the pan global biotechnological concern of an ageing population (Powell and Pearson, 1995), are all symptomatic of the increasing attraction of developing country markets, which are under served domestically.

The EU's BIOMED programme (see table 7.2), has also been an important part of the EU's International Co-operation for Developing Countries (INCO-DC) set of programmes (CEC, 1997). INCO-DC programmes, more-

Table 7.2 Major EU biotechnology programmes (shared cost and concerted actions)

Programme	Year	Objectives	Funding	Comments
Medical Health Research (MHR)	1978–92	Variety of contemporary medicinal projects (over 4 consecutive programmes)	Not available	Acted as lead into the BIOMET I programme.
BIOMED I	1990–94	Pharmaceuticals, occupational and environmental health, Biomedical technology public health, AIDS, TB, infectious diseases, cancer and cardiovascular research	134 mECU	Largely basic research driven.
BIOMET II	1994–98	Brain research, chronic diseases and ageing, human genome research, biomedical ethics	154 mECU[a]	Three foci: • health care providers • large companies • SMEs However, industrial interest remained low. Main interests of project participants were therapeutics, diagnostics and epidemiological.
Biotechnology action programme (BAP)	1985–90	Successor to the biomolecular engineering programme (BEP) (1982–85).	Not available	High industry involvement
Biotechnology Research for Innovation, Development and Growth in Europe (BRIDGE)	1990–94	Information infrastructure, enabling technologies, cellular biology, pre-normative research.	34.5 mECU	High industry involvement
BIOTECH I	1992–94	Molecular biotechnology, cellular biotechnology, ecology and populations, horizontal activities	158.1 mECU	Low industry involvement
BIOTECH II	1994–98	Cell factories, immunology, infrastructures, genome analysis, plant and animal biotechnology, cell communication, structural biology, pre-normative research	73.4 mECU	High industry involvement

Source: CEC, 1997, 1997a, 1997b, DG XII 1998.
Note: [a] First call only.

over, provide key global activities for the EU on issues such as international fish stocks, the Brazilian rainforests, regional organisations in Africa and international organisations in the Association of South East Asian Nations (ASEAN) and Latin America (CEC, 1997b). INCO measures have also helped to found the International Science and Technology Centre (ISTC) in Moscow (1992). All these activities serve to promote EU interests and market access through reciprocity, whilst also ensuring knowledge and technology flows. For example, European agriculture largely derives its plant genetic base from tropical and subtropical species and hence continued access is a major industrial requirement (CEC, 1997b: 29). However, it is worth noting that the attraction of less-stringent regulatory frameworks can encourage less-welcome biotechnology activities that would not be acceptable in developed countries. The recent attack on a US biopharmaceutical plant in the Sudan, over allegations of biotechnology warfare activities, and the continued developing country impression of new biotechnology methodologies imposed by the North adds evidence to increased Southern concerns (Wechsler, 1998: CEC, 1997b: 32)

Within Europe, the Commission communication on industrial policy in 1994 (COM (94) 319 Final) addresses the development of intentional and unintentional industrial power in this sector. For example, extensive reference is made to building and developing increased economic ties with the ACP and CEE states. Indeed in the CEE states, pressure has already been exerted upon the cheaper manufacturers in those countries to limit production, so as to not, unduly, affect the prevailing economic situation within the EU. The lever for this has been future industrial co-operation and the transfer of technology, including biotechnology. The agreement to this effect, has been labelled as the 'Memorandum of Understanding' (COM (94) 319 Final 1994: 23).

The focus of ACP and CEE states for EU industrial policy measures is primarily driven by the emergence of growing markets that can be effectively captured by EU and European producers. These can then be guided to use European standards, through control of the intellectual property regime (IPR) (APEC, 1998: 70). The PHARE programme (Poland and Hungary assistance for economic restructuring – which was extended to other CEE states in 1990), has already sought to develop such objectives, through standardisation and necessary certification for firms doing business in the EU. In those markets where European industries face strong competitive forces, that can affect both employment and production levels, political intervention has been frequently employed, on a bilateral basis, to safeguard the European position (in semiconductors and automobiles for example (Lawton, 1997; Mason, 1994). In other sectors, multilateral discussion has been used but often with only temporary results requiring episodic amendments (in the aeronautics, steel and audio-visual industries for example). Conferences on the international harmonisation of regulatory frameworks have made significant progress in developing a universal framework for new drug applications (NDA), but still reflect regional strengths and weaknesses. Bilateral and multilateral agreements on the recognition of different regional regulatory frameworks remain a significant structural agent, dominated by the developed countries.

Technology trade figures highlight the dependency of the Latin American countries upon imports, mainly from Europe and Asia, and the efforts made by European nations to capitalise upon ex-colonial ties. The countries of Brazil, Chile and Argentina, which possess considerably sized markets, are making concerted efforts to increase their technological standing in the regional environment. Brazil in particular possesses significant human resources at both a state (research institutions) and private sector level. Private sector funding has also increased considerably over the 1990s (their 1994 R&D/ GDP funding ratio was 0.7 per cent in 1994). Brazil has implemented technology transfer initiatives that target information technology, biotechnology and informatics, and is moving to develop these technologies to an internationally competitive standard and quality (OTP, 1997). Cuba, for example, as another developing country with particular biotechnology strengths, has found gaining access to Northern markets difficult for indigenous firms owing to obstacles raised by transnational corporations (Newsedge, 1999e).

Galhardi's (1994) extensive study of NBFs in Brazil supports this perspective, where indigenous firms are proactive in using the research knowledge and skills of regional universities, other indigenous firms and foreign firms. However, the key difference between the NBF in the developing countries and those in the US, or other developed countries, was argued to be in their roles. In the latter, the NBF can operate as a 'research boutique' (Williams, chapter 5, this volume), where the large firm with manufacturing capabilities can use it to source new knowledge, insight and skills. In the developing country, the situation is reversed. The indigenous NBF sources the large firm for new technology and knowledge with production geared to the domestic market. Flow of knowledge and technology is therefore reversed and hence the ability of the transnational co-operation to shape the developing country's biotechnology sector is enhanced.

Chile, as with Brazil, has used its country-specific advantages and natural resources to fund a public science and technology (S&T) programme, which ostensibly guides private industrial development. The current R&D to GDP ratio is approximately 0.8 per cent, but it is expected to rise to 1.0 per cent in 2000. Chile's industrial strengths in agriculture, forestry, astronomy, biology and biotechnology are derived from its regional attributes rather than in the generic enabling technologies. Public policy acts as a source of that knowledge rather than seeking to develop it on a global scale. Similarly, Argentina possesses similar resources, although much of its regional technological strengths are derived from military support and funding (OTP, 1997).

Overall, these considerations place strategic limits on the choices available to the managers of a biotechnological business in the developing and developed country, which will shape the interfaces between industry and government. For the Southern biotechnological firm, technology acquisition (rather than generation) brings codified knowledge to the industry, whilst activities pioneered by the state domestically, and with other governments, increase the overall stock of human capital available for the sector. The routes to knowledge for firms in developing countries are therefore addressed, but with differing industry–state emphasis from those in the developed country.

Biotechnology policy priorities in the market

The issues addressed by Sklair's (1998) first pillar, identifying a global system, include public policy priorities, funding levels, FDI and other structural and relational market-oriented activities. This primarily revolves around the Triad regions as sources of biotechnological knowledge. However, developing-country activities have also been shown to be important and are illustrated as potentially significant actors. In general, the expansion of the biotechnology industry in the Triad between 1992–2000 has regional market sizes of (MITI, 1998):

- US market at $8 billion in 1992 to be $40 billion in 2000;
- Japanese market at $3.8 billion in 1992 to be $35 billion in 2000;
- EU market at $3 billion in 1992 to be $35 billion in 2000.

The global market is therefore estimated to be worth around $100 billion in 2000. Market growth within the EU is given by table 7.3. However, comparative Business Enterprise R&D (BERD) figures between the Triad are illustrative of the significant variance in private investment (see table 7.4). The recent APEC report (1998) stresses, as a counterpoint to the BIOLACs findings (DaSilva and Taylor, 1998), that South–South technologically driven relationships do derive significant benefit when Northern countries' increase R & D investment. That is to say, the South enjoys a technological spillover, although the relative increase is marginal and estimated to be around 6 per cent of the increase in the output of the developing countries.

What is immediately noticeable from table 7.4, are the divergent spending

Table 7.3 The development of the EU biotechnology market

Indicator	1996	1997	1998
R&D expenditure	+21%	+20%	+27%
Number of companies	584 (+20%)	716 (+23%)	1,036 (+45%)
Employees	17,200 (+7%)	27,000 (+60%)	39,045 (+42%)

Source: Thumm, 1999.

Table 7.4 Comparative Triad BERD

Region	Gross Domestic Expenditure on R&D (billions of local currency units)	Percentage of R&D expenditure by sector of performance (1995)	
		Government funding	BERD expenditure
US	178.6	9.2(–)	72.9(=)
China	28.6	44.0(–)	31.9(+)
Korea	9,440.6	3.5(=)	73.1(+)
Thailand	Not available	48.8(–)	7.3(+)
Japan	0.6	10.2(+)	71.2(–)

Source: APEC, 1998.
Note: –,+ and = denote how the share of R&D funding by the actor has changed since 1990.

directions in the US and Japan. Japan is reorienting its spending priorities so as to reflect particular country characteristics, on the basis of the importance of knowledge in technological industries. Government investment is increasing in basic R&D initiatives and facility development, although this is still significantly below the public expenditure in the USA and EU average (NISTEP, 1997).

Health is the largest academic field, in terms of the number of R&D scientists/engineers, within the natural sciences in Japan. In 1997 it underwent a significant reorganisation to emphasise the importance of the biotechnology industry (NISTEP, 1997; Wechsler, 1998). Furthermore, in 1995, registered academic societies were most numerous in the medical and humanities areas in Japan (NISTEP, 1997). It is, therefore, no surprise to add that Japan also has a large share of scientific papers in pharmacology – much higher, for example, than other emerging technological commercial markets, such as earth and environmental science. Japanese consolidation and nation building in this sector is clear (MITI , 1998).

Applying Sklair's (1998) indicators of a global to table 7.4 indicates the shifts in the flow of US FDI R&D funding. Two results are very clear. First, there is a significant increase in Asian investment. Secondly, the comparative increase in US funding is only marginally less than that for the EU. Whilst this is non-technology specific, it does lend anecdotal support to the earlier observation on the emerging markets for knowledge-intensive products in those countries. Between 1983 and 1993, the US significantly increased R&D investment in biotechnology knowledge-rich countries, such as Singapore and Switzerland, whilst also keeping an eye on market access with increased investment in Ireland (as a gateway to the EU), Mexico and Brazil (Dalton and Serapio, 1995).

In the USA, drugs and medicine R&D receive 40 per cent of industrial funds (Dalton and Serapio, 1995). On the export strengths of the major competitive states of the Triad, the recent NSF (1996) review, ranked the USA as top in the biotechnology industries, followed by Germany and then Japan. This pattern was repeated for the life science technologies. However, for the important supporting industries of IT and manufacturing, the lead switched to Japan and Germany respectively.

Following on from the discussion of strategic alliances as knowledge conduits, and as part of Sklair's (1998) global system components, figure 7.3 highlights the variance by country of origin for biotechnology alliances, between the Triad regions. The data in figure 7.3 can be re-plotted to show the preferences in strategic alliance partners for Europe, the US and Japan, which according to the Lawton framework, are indicative of the location of structural power in the industry (Lawton, 1997; IPTS, 1997). Figure 7.4 indicates European preferences, highlighting the increasing trend for EU firms to seek out US partners. Japanese preferred technological collaborative partners, compared with Japanese–Japanese partnering, peaked for Japanese–European firms during the establishment of the single European market (1986–93). Otherwise Japanese–US partnerships are preferred. Notably, US biotechnology firms are actively sought as partners for both Japanese and European companies, whilst the level of US–US alliances has also increased significantly over recent years.

Japanese firms generally have stepped up their overseas business activities since the 1991 collapse of the bubble economy, particularly in Asia (STA, 1998). This was aided by the relaxation of foreign ownership restrictions in Asian countries for both investment and research programme participation (STA, 1998b: BioAsia Monitor, 1998). EU–US collaborative ventures have enjoyed the largest overall increase in collaborative activities and biotechnology collaborative ventures are the primary focus of EU–US inter-

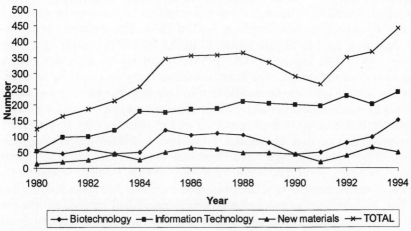

Figure 7.3 Regional Triad preferences for biotechnology strategic alliances (1980–94)
Source: NSF, 1995, Appendix table 4.38.

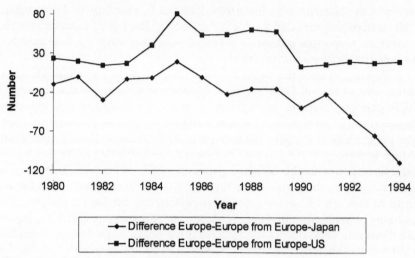

Figure 7.4 Preferred origins of collaborative partners in all technologies for Europe (1980–94)

Note: All technologies include biotechnology, information technology and new materials.

Source: NSF, 1995, Appendix table 4.38.

firm activity. There has been a sharp increase in intra-regional (domestic) collaborative activity in the USA since 1990, which is not reflected in the EU or Japan.

Further insight on the changing nature of the biotechnology sector and the importance of increasing contacts with emergent markets, was highlighted by the 1997 NISTEP report on the level of research support workers in Japan. Since 1994 the level of Japanese research support workers per R&D scientist fell continuously, reaching 0.41 in 1996. This compares with 1.07 in Germany, 1.15 in France and 0.99 in the UK (all European figures are from 1993). The Japanese STA (1998b) report on *'Globalisation and the Japanese Economy'* stresses the broad economic function served by near and far markets for Japanese firms. The former, largely in the proximity of Japan, serves to reduce production costs whilst developing a market base, whilst the latter market is home to both consumer and knowledge, with, in particular, the protection of that knowledge. Furthermore, NISTEP (1997) identifies that 90 per cent of foreign researchers entering Japan in 1995 came from other Asian countries and were study and training focused. However, the level of researchers leaving for developed countries has been falling since 1993, although the number of individuals despatched for cultural reasons to developed countries has been increasing, especially to Central and South America. Understanding the market and cultural issues is arguably a key factor for Japanese firms seeking greater access to the export markets mentioned previously.

A similar pattern is highlighted when we consider the broad investment patterns of German firms. The German FDI share in CEE states shows a small, but rising increase in financial and knowledge flows. From the US standpoint however, investment in Latin America shows no significant variation from its East Asian levels.

Overall, European industry has the lowest presence of the Triad regions (plus the greater economic system, embracing South East Asia, Latin America and CEE states) in both high-technology markets and in high-growth-potential markets (COM (94) 319 Final: 16). Europe also has the lowest presence in terms of its industry growth in those sectors that are growing the fastest. The US by contrast, after suffering a significant general downturn in its technological industries over the 1980s through Japanese and Asian competition, has recovered to the extent that the generic technology industries of information technology and biotechnology are growing at twice the rate of the national economy (Mitchell, 1997). Whilst the EU remains strong in mature technologies, its activity and global market share of the carrier industries (biotechnology, IT and new materials) are comparatively poor (Humbert, 1994).

This has been reflected in the orientation of EU firms towards Japanese and, especially, American biotechnology partners, as well as in the comparatively poor industrial participation rates of both large and small firms in the EU's biotechnology and biomedicine programmes (see table 7.2) (CEC, 1997b). Of more concern is the increased substitute role of EU programmes from a national perspective. Whilst an important but minor reason for participation in the BIOMED I programme was a lack of national funding, in BIOMED II this was a major reason cited by programme participants for application (CEC, 1997b: 16). A similar finding was revealed in BIOTECH I and II programme reviews (CEC, 1997a). These facts, coupled with comparatively small international funding levels, and concerns over the orientation and direction of EU biotechnology programmes (CEC, 1997b), is a continuing difficulty for the EU biotechnology sector.

Biotechnology firms seeking a viable competitive strategy through soft export markets can look to nation-state and regional group pressure, and policy measures. In the external (to the EU) environment, this role has been pursued through political dialogue with the other Triad regions as well as economic action and the explicit threat of future economic difficulties with ACP and CEE states or implicit activities such as the PRAP programme. Internally, this has manifested itself as capacity-reduction programmes rather than purely relying upon market forces to reduce production.

The OECD (1997) identifies reciprocity, through equal programme access in host and home country, as a key consideration guiding foreign firm access in domestic support programmes, with concerns also over transparency and exploitation. Transparency is concerned with clarity and uniformity of guidelines in allowing firms to participate in domestic support programmes. Despite no explicit limits on access to any firm, implicit constraints are active in US, EU and Japanese initiatives. Where commercial exploitation was concerned, initiatives excluded non domestic/domiciled firms unless mutual benefit could be assured through knowledge exchanges and/or manufacture in the home market. However, some industries were identified as being strategic, including the enabling technologies for which a case by case basis applied, unless specific bilateral arrangements had been previously engineered.

Obvious fears over transferring knowledge, intellectual property and competitive advantage to non-domestic firms drive these constraints. However, there was a recognised shift in attitudes between developing and developed country firms, where less-knowledge-based concerns were evident for firms from developed countries, except in the strategically identified sectors. When programmes were strictly confined to domestic firms only, notably in the strategic sectors including biotechnology, three clear situations can be defined. They revolve around a pure internal focus (in this case either nation-state or regional grouping), an expansionist focus (where non home domiciled firms are allowed access potentially subject to reciprocity, transparency and exploitation) and a globally focused initiative which seeks from the start to incorporate all leading industrial actors in an initiative. Such programmes, for example, would aim to develop standards for an industry (see table 7.5).

Table 7.5 Matrix of regionally focused biotechnology initiativities and objectives

Collaborative focus	Activity	Objectives
Internal to the country (domestic firms)	Consolidation (examples include VLSI, EUREKA, early SEMATECH, JESSI, ES^2)	Developing internal resources to industry competitive level
External to the country (all firms)	Expansion (examples include ESPRIT phase III/IV, late SEMATECH)	New to the domestic group knowledge and skills
Globally focused	Standard setting	Standard setting

Notes: VLSI – The Japanese semiconductor programme of the 1980s – the very large scale initiative

EUREKA – French inspired pan European technological collaborative programme (1985)

SEMATECH – US semiconductor research consortium – semiconductor manufacturing and technology institute

JESSI – The Joint European Sub micron Silicon Initiative (Siemens, Phillips, GEC, SGS Thomson, STET)

ES^2 – Joint venture of Olivetti, Bull, Phillips, GEC, Siemens in the early 1980s (matched by US^2 in the US)

ESPRIT – The European Strategic Programme for Research in Information Technology (since 1983).

Conclusion

This chapter has attempted to outline the benefits of using a particular IPE approach to study the biotechnological sector. It began by exploring conceptually how contemporary thought on technological industries can be accommodated within the IPE perspective, which particularly focuses upon knowledge as a key determinant of competitive success. Non-economic and societal factors were argued to be important considerations, which, in conjunction with historical strengths, can be used to understand why and how the demands of the biotechnology sector fit better in one country than another. Shifts in Japan, in terms of consolidation, knowledge generation and appropriability, and the developing countries with their different perspectives on South development and the use of transnational enterprises from the developed countries, were highlighted as examples. Unique market characteristics, however, have necessitated a more mixed approach to policy development underpinned by both economic and non-economic factors.

Overall, within the EU, the need to address at least three different types of biotechnology firm, which have different knowledge, capital and risk requirements, arguably results in different policy demands not presently being accommodated, despite current strengths. In addition, the influence of three-tiered clinical and regulatory environments, and the effective protection of intellectual property are redistributing sources of structural influence in this sector, largely from the EU to the US. However, cautious optimism remains within the EU largely because of its embedded knowledge infrastructure that can support the different types of biotechnology firm emerging. In essence, whilst fragmentation of the European market continues to pose obstacles to competitive development, it can also be argued that this ensures EU firms are able to access policy measures that do address different types of biotechnology firm which may not be so easily undertaken in more homogenous markets.

By using the first Sklair (1998) criteria for a global system, information was presented on recent biotechnology initiatives, funding levels, priorities and knowledge dimensions across both developed and developing countries, to illustrate the different perspectives of competitive strength in this sector. Notably, by considering access to these public policy initiatives two general themes emerged. Developed country policy priorities (public and private) are concerned with biotechnology firm types, capital, the IPR and the structural market characteristics between different developed country actors. For developing country actors, the focus shifted and became one of market development and structural control (examples were given of Japanese and EU activities in particular).

In terms of the Lawton (1997) IPE framework, firm–EU Commission relationships comprise pre-competitive and industrial research collaboration. Access is maintained for all EU firms but with limited *ad hoc* access to non–EU domiciled firms. Access to developing country markets and bilateral agreements shape market development. EU member state and Commission relationships focus upon the development of European perspectives for the biotechnology sector, including IPR, healthcare reforms and funding measures. They intensively engage in global collaborative activities. On a nation-state to nation-state level, technology and knowledge flows are shaped by reciprocity, transparency and mutual exploitation. However, on the nation-state to firm level, basic research input and development is the key activity, but in the context of changing societal values of the nation-state. Finally, from a firm-to-firm perspective, internal and external collaborative activities are evident, shaped by developed country knowledge flows and developing country firms' market and technology conduit activities. Overall:

- There is a shift in developed country focus from predominently research activities to applied, industry development initiatives.
- There are shifts in the focus of some developing country funding priorities to basic research and human capital development, as a recognition of the knowledge-demanding nature of the biotechnological sector.
- There is the emergence of a range of different types of biotechnology firms and NBFs which place different and opposing demands upon the state as supplier and consumer of biotechnological knowledge and products.
- There is a pressing need to incorporate socio-economic dynamics which enable a better fit between the biotechnology industry demands and mechanisms of a given market.

8

An international environmental regime for biotechnology

John Vogler and Désirée McGraw

Biotechnology could pose a potentially catastrophic threat to the natural environment. It may also yield very significant benefits in terms of the restriction of pesticide use and restoration of habitats. At the moment it is simply impossible to know with any certainty. The question is becoming urgent as transgenic organisms move out of the laboratory and into the global marketplace. The new biotechnology is subversive of the existing regulatory architecture for food safety and seed certification. Acute uncertainty associated with the effects of genetic engineering, when organisms are capable of autonomous growth, makes its products an obvious candidate for the application of the 'precautionary principle'. Yet such an approach is disallowed under the current GATT/WTO regime for international trade.

What has been the response of the international community to the manifold and uncertain environmental implications of the new biotechnology? At a formal level, it has been limited to the negotiation of a regime for 'biosafety' in relation to 'living modified organisms' (LMOs) transferred or traded across international boundaries. The construction of a regime, in the form of a Biosafety Protocol to the 1992 Convention on Biological Diversity (CBD), proceeded largely unnoticed until the negotiations, held in Cartagena, Colombia, collapsed in February 1999. The impasse occurred amidst rising public concern about the implications of genetically modified (GM) products and rumours of a trade war between the United States, convinced of the benign and beneficial character of its genetically engineered agricultural exports, and an increasingly ambivalent and even environmentally protectionist European Union. Strenuous attempts were made during the following year to revive the Protocol, and in the final days of January 2000 a text was agreed at a resumed Extraordinary meeting of the Conference of the Parties to the CBD in Montreal.

This chapter uses a regime analysis framework to consider the characteristics of the emergent biosafety regime and the issues and stakes involved in its prolonged negotiation. Before doing so, it raises the broader question of why international co-operation might be required to deal with the potential environmental threats posed by the new biotechnology. Regimes are governance institutions at the international level around which actor expectations and behaviour converge. They govern specific issue-areas. Adopting a regime perspective alerts us to the ways in which issue-areas are constructed. In this case, it highlights the question of how biosafety became the focus of regime creation at the expense of other potential issue-areas which might well be regarded as having much greater significance in terms of biodiversity conservation. A regime perspective also throws into relief the fragmented nature of governance institutions at the international level. Here the significant divide is between attempts to build regimes for environmental protection and the well-established and dominant regime for international trade.

Biotechnology as an environmental problem

The question of the environmental impact of biotechnology has been inextricably connected with debates about biodiversity and species loss, which found their way on to the international agenda after the first UN Conference on the Human Environment held at Stockholm in 1972. From then on, the conservation of genetic diversity became one of the principal activities of the newly established UN Environment Programme (UNEP).

A global biodiversity convention was first proposed by the International Union for the Conservation of Nature (IUCN)[1] and from 1987 appeared on the formal agenda of UNEP. In the same year, biotechnology was mentioned in the celebrated Brundtland Report (WCED, 1987) in terms of the central ambiguity that has characterised most discussion. Here was a technology that was truly *double-edged* in terms of its environmental implications. On the one hand, there is a *promethean* (Dryzek, 1997) enthusiasm that biotechnology could 'yield cleaner and more efficient alternatives to many wasteful processes and polluting products' (WCED, 1987: 218). Yet, on the other hand, as with other new technologies, it is not intrinsically benign: 'New life forms produced by genetic engineering should be carefully tested and assessed for their potential impact on health and on the maintenance of genetic diversity and ecological balance before they are introduced to the market and thus to the environment' (ibid.: 219).

Many NGOs and campaigning groups, aware of public distrust and alarm over GM foods, have tended to portray biotechnology and biotechnically engineered products as a potential public health and ecological disaster. This is more than a tactical ploy, it represents a rejection of the rationalistic 'value expectations' approach prevalent in the scientific and policy communities, according to which it remains an open and empirical question as to whether biotechnology can prove to be environmentally beneficial. To use Dryzek's categories once again, the rejectionist discourse is that of 'green romanticism'. Put simply, from this perspective biotechnology is not evaluated in terms of its potential consequences, but rejected *in toto* as the ultimate anthropocentric assault upon a sacred natural order. Scientific rationality and technical manipulation (with its atomistic conceptions and separation of humankind from its ecological context) have, in this view, been the core problems since the days of Francis Bacon.

The sceptic might ask why biodiversity was considered a global environmental problem at all. It does not, after all, have the status of a 'global commons' or a common resource beyond sovereign jurisdiction. Its components are largely located within national jurisdictions. Nevertheless, as with forests, the longer-term consequences of depletion and degradation have implications for the whole biosphere. Swanson (1999) has argued that the existence of the CBD as a global convention has an underlying rationale in terms of collective land-use planning, which cannot be achieved on a national basis, through the 'creation of incentive mechanisms that will translate the need for designated reserves and resources into their effective maintenance and management' (ibid.: 331). Although the biotechnology industry is clearly central to the latter endeavour, a protocol on biosafety would not appear to be the most immediately relevant international instrument to achieve such aims. It is debatable whether the expenditure of so much effort in creating a protocol in such a difficult and contentious area constituted the best way of developing the potential of the CBD. Questions might also be asked as to whether the 'living modified organisms' (LMOs), which the Protocol aims to control, represent a primary threat to biodiversity.[2] Despite the fact that there are much less-contested threats, such as the introduction of alien species, the decision-makers within the CBD chose to give priority to biosafety and LMOs – the products of the biotechnology industry.

One of the most evident ways in which the biotechnology issue is different from analogous international environmental concerns is that – despite the specific uncertainties of the science and policy communities as well as the blanket rejectionism of much of the environmental movement – it lacks a clearly defined *environmental problematique*. The contrast is strong even with the climate change issue, where modelling and prediction has been hotly disputed. This is clearly one reason why it is easier to identify biotechnology as a human health or trade problem rather than an environmental threat. From an environmental perspective, the most immediately identifiable problem – and one which the Biosafety Protocol seeks to address – is the danger that LMOs may have unintended environmental consequences when *released*. The image is of a genetically engineered seed that might run amok disrupting and destroying ecosystems. For twenty years, national and EU regulatory activity has been concerned with estimating the risks associated with the release of new products, and with the safe containment of research and development. The controversial farm-scale trials of GM crops, underway in the United Kingdom from 1999 to 2004, were designed to establish their environmental impact before the licensing of seed for commercial use.

There are various mechanisms whereby environmental harm might be caused. There may be interference with wider ecological processes (this appears to have been the case with 'ice-minus' experiments intended to reduce ice formation on crops). Genetically modified plants are often designed to resist particular herbicides (particularly through the incorporation of the Bt gene), but problems may arise if herbicide-tolerant genes transfer to other plants, thereby raising the possibility of the production of superweeds. At the same time, the chain of consequences associated with even small alterations in an ecosystem may be long and complex, and GM plants have been associated with reductions, for example, in bird and insect populations. The inevitably restricted temporal and spatial scope of current trials, compounded by the problems of replication and the possibility that gene development will follow an unforeseen vector, mean that existing methodologies cannot claim to provide comprehensive reassurances on environmental impact. To argue otherwise is to risk further damage to the public authority of 'official science' in this area (Wynne, 1999).

Once GM products are traded across borders, or even if GM modified crops are grown adjacent to national frontiers, the argument for international co-operation is clear. Those who are suspicious of current scientific environmental impact assessment and associated policies, would argue (as does Greenpeace, for example) that this should take the form of a moratorium. Alternatively, if the validity of at least some existing testing regimes is accepted, then the incorporation of national regulatory activities into a universal system embodying 'best practice' becomes necessary. The analogy would be with various international pollution control regimes which attempt to bring national abatement policies into line. Such a regime might well provide the solution to a 'prisoners' dilemma' or 'tragedy of the commons' type problem posed by manufacturers who take safety measures to protect the environment, for example by mixing GM seed with normal seed in their products (Thompson 1997: 106). Safety requirements would probably be imposed by national governments. However, outside their jurisdiction firms might have short-term individualistic incentives to produce cheaper and more effective products that put the environment at risk. It would then be in the interest of the majority of firms and governments to promote and enforce strong international standards to protect both their commercial interests and the environment. This case has analogies with the Montreal Protocol on Ozone Depleting Substances and the incentives for leading producers of chlorofluorocarbon (CFC) alternatives to participate in and extend the regime which phased-out the production and consumption of CFCs. Allowing the continued production and export of CFCs anywhere in the world would have placed those who had been forced to invest in more expensive CFC alternatives at a competitive disadvantage.

The problem with such arguments by analogy is that, in comparison with existing pollution control regimes, little is known about the specific environmental effects of LMOs. Moreover, countries do not necessarily know whether the products they are importing contain LMOs or not. Such uncertainty leads to difficulties in making accurate cost–benefit calculations. A blanket application of the precautionary principle (which would place a moratorium on commercial biotechnology development and/or trade) is clearly unlikely at a national level and, despite some of the demands of environmental activists, internationally non-negotiable. An important existing dispute arises from US objections to the complex and 'precautionary' nature of EU approval procedures, in place since 1990, for the 'release' of GMOs. An evident distinction between the biotechnology and, for example, stratospheric ozone and climate change issue-areas is that there is nothing at present which can be either banned (as CFCs under the Montreal Protocol) or reduced (as the six greenhouse gases under the Kyoto Protocol). Instead, as has already been noted in this book, LMOs have to be treated on a 'case-by-case' basis.

The previous discussion has centered upon the possibilities for international co-operation in coping with the unintended environmental consequences of generally valued activities. Yet, it is arguable that the *intended* consequences of such biotechnology activities pose the greatest environmental threat – through the production of monocultures and the attendant reduction of biodiversity (Weizacker, 1994). This may render crops and perhaps whole ecosystems liable to collapse and set up complex feedbacks which diminish the possibility of sustainable development. The monoculture problem does, of course, pre-date modern biotechnology. Indeed, there is no necessary connection, except that biotechnology allows the possibility of greater speed and efficiency than previous plant-breeding methods.

It has been argued that the idea of valuing and protecting the natural environment against human intervention is a chimera. Millennia of human interference have rendered the very idea of the 'natural' non-sensical, and total re-engineering of the environment is both inevitable and desirable (Sagoff, 1991). Nonetheless, a consensus exists on the need to conserve biodiversity – if not for its intrinsic value at least for its potential role in alleviating possible monocultural disasters, brought about by the application of biotechnology. Equally, biotechnology itself does not create biodiversity; rather, it relies for its very existence on the re-ordering of existing genetic material. In the language of the CBD, '"Biotechnology" means any technological application that uses biological systems, living organisms, or derivatives thereof, to make or modify products or processes for specific use' (Article 2).

Because as yet unexploited genetic material is universally valuable, yet located *in situ* at particular locations (usually in the developing world and especially in threatened tropical rain forests), there is an argument for creating a common heritage conservation regime for a global environmental resource.[3] The concept of the *Common Heritage of Mankind* (CHM) was first mooted in discussions preceding the Third UN Law of the Sea Conference in the late 1960s. It proved to be highly controversial in discussions of the principles for distributing the mineral resources of the deep seabed and was applied, with only partial success, to the allocation of positions for satellites in geostationary orbit (GSO) (Vogler, 1995: 32–5). CHM as a regime principle was very much part of the campaign for a New International Economic Order waged by the developing countries in the 1970s. It implies that (usually non-living) resources 'cannot be appropriated to the exclusive sovereignty of states but must be conserved and exploited for the benefit of all without discrimination' (Birnie and Boyle, 1992: 120). While this was very much in line with the interests of developing countries with respect to resources that were in danger of being monopolised by the industrialised world, the position was essentially reversed in respect of genetic resources. Industrialised countries had always been able to enjoy unrestricted access to the genetic riches of the South, and a common heritage regime proved to obstruct attempts by developing countries to control and extract revenue from genetic resources located within their territory.

The CBD employs the more enigmatic 'common concern of humankind' – a formulation which avoids the idea that biodiversity or genetic resources should be accorded 'common heritage' status. Instead, throughout the development of the CBD, developing countries 're-affirmed' their national economic sovereignty over such resources. The innovation of the CBD was to provide for the 'fair and equitable sharing of benefits arising out of the utilization of genetic resources' as one of its three major objectives (alongside biodiversity conservation and sustainable use of its components). This 'benefit-sharing' would occur in exchange for access to biological resources which, unlike the deep seabed or GSO, were located within state boundaries.

There are still those who advocate a form of common heritage regime. Their motivation is provided by observation of the activities of biotechnology companies and governments in the present liberal international economic order:

> There is another way. Turn our backs on the present European policy of allowing life forms to be patented, and declare living things 'the common property of humanity'. And reorganise genuinely public research around this common property in order to block the already well-advanced private hold that is seeking to eliminate any scientific alternative that would make ecologically responsible and sustainable agriculture possible. Guarantee the free movement of knowledge and genetic resources that have made the extraordinary advances of the last 60 years possible. Restore power over living things to the farmers, that is to each one of us. Replace economic warfare and the plundering of genetic resources with international cooperation and peace. (Jean Pierre

Berlan and Richard C. Lewontin 1999: *Le Monde Diplomatique*, 29 January, reprinted in *Guardian* 22 February.)

For the present, all this remains in the realm of utopia. At the international level, the biotechnology issue has been formally addressed under the CBD through the negotiation of a narrowly defined 'Biosafety Protocol' for the transboundary movement of LMOs.

The biosafety issue-area

Issue-areas, as Keohane (1984: 61) reminds us, are 'sets of issues that are in fact dealt with in common negotiations and by the same and closely coordinated bureaucracies'. The legal and organisational context within which the biosafety issue-area has been framed has been provided by the CBD. In line with contemporary practice in environmental treaty making, the Convention provides both an international expression of concern and a set of arrangements under which scientific efforts can be mobilised to provide a clear and consensual analysis of the problem, which will serve as the basis for specific regulatory control through the subsequent development of protocols. The Biosafety Protocol represents the first attempt to operationalise a key and contentious part of the CBD. If it proves to be successful, it may become – as was the case with the Montreal Protocol in relation to the Vienna Convention (on substances that deplete the stratospheric ozone layer) – more significant and better known than its mother treaty. Indeed, the conclusion of the Biosafety Protocol was largely framed as a litmus test for the credibility of the Convention. Given the contentiousness of the issue-area, selecting biosafety as the focus of a first protocol under the CBD was a high-risk political strategy on which to stake the future of the Convention.

From its inception the CBD has had a very bad press (Brenton, 1994: 204). Arising from an inchoate concern with species loss, as well as the legal initiative of the IUCN, its development was swept along in a tide of pre-Rio enthusiasm. It became inextricably connected in various ways with the parallel negotiation of the Framework Convention on Climate Change (FCCC) and other aspects of the 1992 Rio Earth Summit.[4] Its negotiation, under the auspices of UNEP in Nairobi, was dominated by North–South anatagonism and a desire by the latter to make up ground that had been lost in the climate change and other multilateral negotiations.[5] Developing countries pressed for equitable compensation for the use of their genetic resources while the US, in particular, urged for a convention that 'concentrated on the protection of biodiversity in species-rich areas without impeding the growth of the biotechnology industry' (McConnell, 1996: 54). The resulting compromise text determined that the Biodiversity Convention would be generally vague, voluntaristic and in advance of any credible scientific underpinning (Munson, 1993; Sanchez and Juma, 1994; Raustiala and Victor, 1996). In the end, the US refused to sign the CBD and, although the Clinton Administration reversed this decision, the Convention has remained unratified by the US Senate.[6]

On one issue, at least, the CBD was clear. Article 19.3 contains a specific mention of a 'control protocol' relating to the safety of 'any living modified organism resulting from biotechnology that may have adverse effects on the conservation and sustainable use of biological diversity'.

The definition of the issue-area and the development of a protocol were carried forward by the first and second meetings of the CBD's Conference of the Parties (COP), held in Nassau in late 1994 and Jakarta in November 1995, respectively. The permanent CBD Secretariat, based in Montreal, initiated a study designed to investigate where a protocol could usefully fill gaps in the existing regulatory system for the transboundary movement of LMOs. It surveyed the various voluntary codes of conduct devised by UNIDO and the OECD, along with older regimes for plant and animal health. The study concluded that 'the existence of gaps presupposes the existence of a system' but 'no such system exists'. For existing international regulatory agreements any coverage of the transboundary movement of LMOs was 'purely incidental to their main purposes' (CBD Secretariat, 1997: 17–18).

The Jakarta COP of 1995 set up a negotiating process for a protocol. However, in doing so, COP-2 became embroiled in the question of the domain of any future regime. This question remained a divisive North–South issue, which was essentially fudged in the negotiating mandate handed down from the COP in Decision II/5. Thus, many industrialised countries and agricultural exporters required a narrow interpretation of the application of a protocol. This would restrict the instrument to the regulation of the transboundary transfer of LMOs (such as seeds, products for field testing and fish) intended for direct release into the environment and directed towards their impact upon biodiversity – strictly defined in terms of flora and fauna. Others, particularly from the South, wanted a wider application which at least addressed 'handling and use' (*ENB*, 1999: September).

The compromise wording of the negotiating mandate (COP Decision 11/5) failed to resolve the question of the future regime's domain. Instead, as negotiations proceeded, references to LMOs and 'products thereof', as well as biodiversity and 'health and socio-economic considerations', were inserted into the draft negotiating text but surrounded by square brackets – indicating the disagreement of some participants. Extending the domain to 'products thereof' would greatly widen the scope of regulation, possibly to include pharmaceuticals and food products which had been bio-engineered. Similarly, widening the grounds upon which Advance Informed Approval (AIA) and risk assessment would be based to include health and socio-economic considerations would provide, in the view of industrialised country exporters, an almost open-ended justification for prohibiting imports. In the

US view, the insertion of these clauses by developing countries represented an attempt to overturn the mandate agreed at Jakarta, rendering them essentially non-negotiable (Rafe Pomerance, Head of US Delegation to Ex-COP1, Press Conference, 19 February 1999).

At the Jakarta COP, the task of producing the text of a biosafety protocol was assigned to an interim negotiating body with the less than snappy title of the Open-Ended *Ad Hoc* Working Group on Biosafety (BSWG). This body met on five occasions prior to 1999 for one-to-two-week sessions. By the end of BSWG-5 at Montreal in August 1998, the negotiating text for a protocol had been refined down to an eighty-page document – half its original length (UNEP/CBD/BSWG/6/8). The text remained replete with brackets, however, reflecting clear disagreement on most of the operational articles. Indeed, thirteen of the forty-two draft articles were bracketed in their entirety. Slow progress at the BSWG had already required the COP, at its fourth meeting at Bratislava in May 1998, to extend the previous schedule for a completed protocol from late 1998 to early 1999. Thus it was that the sixth meeting of the BSWG was arranged for February 1999 in Cartagena, Colombia, to immediately precede an extraordinary meeting of the COP intended to provide ministerial approval of the completed handiwork of the BSWG negotiators.

In the event, such expectations were premature. Although it was agreed that the protocol would be named the 'Cartagena Protocol', there was no final resolution of the difficulties that had continued to dog the negotiations on biosafety. These, broadly, pitted the majority of developing world (the so-called "Like-Minded Group") against those industrialised countries with biotechnology industries and agricultural exporters (the latter gathered in the Miami Group comprised of Argentina, Australia, Canada, Chile, Uruguay, and the US) with the European Union acting in a mediating capacity.[7] Whereas the original negotiations on the CBD could be essentially understood in North–South terms, the biosafety negotiations began to take on a more complex character as countries with nascent biotech industries, or with interests in large-scale agricultural exporting, reconsidered their interests and alignments. A specific example is provided by Argentina who parted company with the Latin American and Caribbean Group at the 1996 BSWG-1.

The events that followed have almost become a ritualistic pattern on these occasions. Round-the-clock informal sessions as the deadline approached yielded a compromise text agreed between the EU and the developing countries. However, this proposal was finally rejected by the Miami Group and the Cartagena meeting broke up in acrimony.

Failure to reach agreement in Cartagena was a major setback, not only for the prospects of concluding a Biosafety Protocol, but also for giving effective content to the CBD itself. Under the chairmanship of the Colombian Environment Minister, Mr Juan Mayr, many efforts were deployed in order to bring the process back on track. Informal consultations took place twice in 1999, first in Montreal (1 July) and later in Vienna (15–19 September), where the Parties agreed on a methodology to deal with the remaining issues, and to carry the process forward to a successful conclusion at the resumed formal session scheduled for 24–28 January in Montreal.

Notwithstanding these constructive efforts, most delegates arriving in Montreal gave the Protocol a 50/50 chance of being adopted. Formal discussions took place in the 'Vienna Setting' – a roundtable chaired by Minister Mayr, with each group being represented by two spokespersons, the rest of the delegates being seated behind. All governments as well as accredited NGOs, industry representatives and media had full access to these deliberations.

By setting up sub-groups with well-defined mandates, Minister Mayr managed to isolate the key issues and remove technical stumbling blocks before for the high-level political process that occurred during the last 24 hours of the meeting. This negotiation process not only cleared the way for the Ministers to make purely political decisions, it produced a general feeling that the negotiations had come too far to fail . Moreover, this structure dispelled the possible fall-back position of any coalition (notably, the Miami Group) that might claim there was not enough time in the end to conclude the Protocol. After difficulties raised at the last minute by the Miami Group in connection with the Protocol rules on accompanying documentation (overcome by discussions with the EU) the text of the Cartagena Protocol on Biosafety was finally adopted by consensus shortly before 5:00 a.m. on Saturday, 29 January.

The Cartagena Protocol on Biosafety

Principles and norms

Principles are defined in the regime literature as 'beliefs of fact, causation and rectitude' (Krasner, 1983). Any discussion of the principles of the biosafety regime must be prefaced with a reference to the underlying difficulty of establishing a scientifically grounded consensus on the dimensions of the problem. The contrast with other environmental regimes, founded upon a common recognition of a harm to be avoided or pollution to be abated, is clear. Almost all established regimes have required a fairly high level of scientific and political consensus that a problem exists, even if its precise dimensions may be disputed and its assumptions later invalidated. Epistemic and cognitive theories (Haas, 1990) isolate this as the primary variable in their explanations of environmental regime creation. In the case of LMOs and biodiversity, no such epistemic consensus exists. Scientific opinion is divided, but would probably tend towards minimising the inherent risks in terms of health or environmental modification, while stigmatising aroused public opinion and NGOs as irrational and alarmist. The CBD, as we have seen, does not define biodiversity as a 'common heritage of mankind', neither does it follow the FCCC in enunciating the 'common but differentiated responsibilities' of industrialised and developing countries. Nonetheless the CBD, taken as a whole, does entail a pattern of separate rights and responsibilities (McGraw, 2000). The problem, as ever, lies in defining and operationalising such principles. Thus, the BSWG meetings were marked by North–South disputes over the duties and liabilities of biotechnology-exporting and biodiverse-importing states.

Most fundamentally, there was a dispute about the purpose of a biosafety regime. Some idea of its extent can be gleaned from the heavily bracketed text relating to the 'Objective' (Article 1) of the draft negotiating text before BSWG-6:

> The objective of this Protocol is [[, in accordance with the precautionary principle,] to contribute to ensuring an adequate level of protection in the field of] [the safe transfer, handling and use [in a transboundary context] [specifically focusing on]] [the safe] [transboundary movement] of living modified organisms [and products thereof] resulting from modern biotechnology that may have adverse effects on the conservation and sustainable use of biological diversity [taking also into account risks to human health [and socio-economic imperatives]].

Behind the brackets lay fundamental differences of principle about the purposes of a biosafety regime. In surveying existing regimes, it is usually clear whether their intent is either trade promotion with environmental safeguards or environmental conservation, perhaps through the use of some trade-related instruments. The Sanitary and Phytosanitary code of the WTO or Art. XX (g) of the GATT fit into the first category, while CITES and the trade provisions of the Montreal Protocol fall into the latter. In the case of the CBD, signatories genuinely disagreed as to the character of the Biosafety Protocol. Was it, on the one hand, a trade facilitation agreement rationalising national safety procedures on an international basis so that all may benefit from a burgeoning trade in bio-engineered seeds and products in a market that would not be disrupted by trade barriers based on spurious or non-existent scientific assessment of the minimal risks involved? Or was it, on the other, a genuine attempt to provide a comprehensive environmental regime affording a generalised high level of protection to habitats and human beings against dangers which could, as yet, hardly be guessed at?

At Cartagena, the Miami Group of agricultural exporters pressed for a regime that would minimise trade barriers leaving approximately 90 per cent of their biotechnology exports untouched by environmental safety legislation. A number of developing countries and NGOs, suspicious of the activities of Monsanto and other agritech transnationals, pressed for more rigorous safeguards, prioritising environmental health and social concerns over those of commerce in biotechnology. In its role as bridge-builder but also, perhaps, because of serious divisions among its Member States, the EU pursued 'a balanced approach' to the Protocol, focusing on the elaboration of a meaningful advanced informed agreement (AIA) procedure, and maintaining that the Protocol must strengthen environmental protection while not necessarily hindering trade.

Relationship with the WTO

The issue emerged most clearly in a dispute about the relationship between the Protocol and the WTO – one of the main causes of the Cartagena debacle and again the subject of heated debate in Montreal. The two opposing views focused on which rules would apply in case of conflict: either the rules set out by the Protocol (a position strongly defended by the EU and the Like-Minded Group) or the rules set out by the WTO (a position vigorously defended by the Miami Group).[8] In Montreal, the Chair's proposed preambular language, containing the concepts of 'mutual supportiveness' and 'equal status' between the Protocol and other international agreements, was in turn rejected by the Miami Group.

Ultimately, the negotiators accepted compromise text:

Recognising that trade and environment agreements should be mutually supportive with a view to achieving sustainable development;
Emphasising that this Protocol shall not be interpreted as implying a change in the rights and obligations of a Party under any existing international agreements;
Understanding that the above recital is not intended to subordinate this Protocol to other international agreements;

The third preambular paragraph effectively neutralises the second, therefore leaving the emphasis on mutual supportiveness. WTO rights and obligations are fully preserved, meaning that Parties to both sets of rules should implement them in a consistent and complementary way.

Precautionary principle

In regime analysis, if not everyday usage, the precautionary principle constitutes a norm. It urges that preventive action to protect the environment need not be based upon full scientific certainty. The principle is enshrined in the Rio Declaration on Environment and Development, and found its way into COP Decision II/5 as well as earlier drafts of Biosafety Protocol. However, it enjoys questionable status in international (environmental) law (Tinker, 1998: 424–5 but see also Sands, 1993: 207). By the Cartagena meeting, references in the negotiating text had been changed to the somewhat weaker 'precautionary approach'. The Miami Group particularly objected to its inclusion in the operational articles of the draft negotiating text on the grounds that these might be invoked by Parties as a means of prohibiting imports in the absence of scientific evidence of their biodiversity damaging effects. The ultimate application of the precautionary approach/principle would be the kind of blanket moratorium on the transfer of LMOs, pending full scientific investigation, favoured by many NGOs.

In Montreal, both the European Union, which had since 1990 adhered to the precautionary principle in its own procedures governing releases, and the Like-Minded Group insisted on its inclusion in the body of the agreement as well as the preamble. As a result of trade-offs on other issues with the Miami Group, this objective was largely achieved. It should be stressed, however, that the operational provisions allowing Parties to take protective measures when faced with scientific uncertainty comes into play only once a *science-based* risk assessment has been carried out. The final text in the operative provisions (Article 10.6) was developed on the basis of a proposal from the Compromise Group. This reads:

Lack of scientific certainty due to insufficient relevant scientific information and knowledge regarding the extent of the potential adverse effects of a living modified organism on the conservation and sustainable use of biological diversity in the Party of import, taking also into account risks to human health, shall not prevent that Party from taking a decision, as appropriate, with regard to the import of the living modified organism in question in order to avoid or minimise such potential adverse effects.

The inclusion of the precautionary approach/principle in a substantive provision of a broadly supported MEA marks an important development in international law

Scope

The precautionary principle was negotiated as part of a package which limited the scope of the Protocol in order to be made acceptable to the Miami Group. The final agreement reflects a compromise between the Like-Minded Group (arguing for a broader scope) and the Miami Group (advocating a limited scope). The Protocol's scope is narrowed in a number of ways: it excludes pharmaceuticals (Article 5) and although LMOs destined for contained use and the transit of LMOs are covered by the Protocol's general provisions, they are excluded from its procedural provisions regarding AIA (Article 6). Most crucially for the Miami Group, the AIA procedures cannot be applied to block the economically very significant import of commodities – LMOs for use as 'food, feed or for processing' (LMO-FFPs). In the text agreed at Montreal, the latter are explicitly subject to a new alternative and less onerous procedure (Article 11, Annex III) involving notification and labelling by exporters.

Rules and decision-making procedures

The Cartagena Protocol adopts the administrative arrangements of its mother treaty, the CBD. For anyone familiar with the organisational dimension of other recent UN environmental treaties, it is clear that the CBD adopts what has become a standard form. The supreme decision-making body is the Conference of the Parties (COP). It usually operates by consensus but can, in the last resort, utilise a two-thirds majority voting system for amendments to the Convention or for the adoption of protocols (CBD Article 29). It would also serve as the authoritative Meeting of the Parties for a Protocol (COP/MOP). To date, the COP's main advisory body[9] is the Subsidiary Body for Scientific, Technical and Technological Advice (SBSTTA in CBD Article 25), which is supposed to function as a multidisciplinary expert source of advice for the COP. Moreover, the CBD, like the FCCC, utilises the Global Environment Facility (GEF) as its financial mechanism, but the relevant negotiations prompted much greater controversy between North and South. It was this issue that was still unresolved at the eleventh hour before the conclusion of the CBD.

Procedure for transboundary movements of LMOs (Articles 7–10, 12)
The general intention of the Biosafety Protocol is to set up procedures allowing national authorities to assess and possibly refuse the transboundary movement of LMOs into their territory. The heart of the system is the Advance Informed Agreement (AIA) procedure. Article 7 of the Protocol sets out differentiated approaches to LMOs for intentional introduction into the environment (such as seeds and plants) and LMOs for direct use as food, feed or for processing (LMO-FFPs). The AIA procedure (Articles 8, 9, 10 and 12) applies to the transboundary movement of LMOs in the former category and specifies comprehensive rules regarding notification, acknowledgement of receipt and decision making. The responsibility for the notification is on the exporter. The decision-making procedure is based on the 'no consent, no movement' principle, protecting the 'sovereign right' of countries to refuse imports of LMOs considered to be a risk to biodiversity. The risk assessment must be science based, with a formulation of the precautionary principle in case of scientific uncertainty.

Alternative approach to LMO commodities (Articles 11–12)
The Protocol provides for an alternative approach to LMO commodities (Articles 11-12). LMOs intended for direct use as food or feed, or for processing (LMO-FFPs) account for roughly 90 per cent of all the value generated by trade in LMOs. Arguing that the transboundary movements of these LMO commodities do not pose a threat to biological diversity, the Miami Group sought to exclude them from the Protocol's scope. However, when this proposal was rejected by all the other groups, the Chair sought a new basis for negotiations in Montreal: an alternative approach to LMO-FFPs which takes into account the specific characteristics of trade in agricultural commodities. This alternative approach, articulated in Articles 11 and 12, creates a system for advance warning about possible imports

through an international information-sharing mechanism – the Biosafety Clearing-House.

Having carefully avoided any reference to words such as 'procedure' and 'notification', the final text of Article 11 was acceptable to the Miami Group because, in its view, the article does not set out a specific Protocol procedure for the import of LMO commodities; it merely provides *guidance* for decision-making by those Parties who do not have in place a domestic regulatory framework (currently, only about sixty Parties have enacted this type of legislation). The other major negotiating groups were able to agree to the Article because it reflected a number of their own objectives, most notably that: transboundary movements of LMO commodities are within the scope of the Protocol; imports of LMO commodities are not forced upon developing countries or others who do not want them but who do not have in place an appropriate domestic regulatory framework; and the role of the precautionary approach in decision making by the Party of import is recognised. In addition, the need for financial and technical assistance and capacity building for developing countries is underlined.

Handling, transport, packaging and identification (Article 18)

The Protocol also takes a differentiated approach regarding handling, transport, packaging and identification of LMOs (Article 18), establishing different requirements for LMOs intended for direct introduction into the environment, LMOs destined for contained use and LMO commodities. Shipments containing *LMO-FFPs* have to be clearly identified as 'may contain' LMOs, but no threshold is specified (Article 18.2a). In view of the interest of all countries in this subject, it was agreed that the Parties to the Protocol would decide on detailed requirements for this purpose, including specification of the LMO-FFP's identity, no later than two years after the Protocol's entry into force. These requirements do not legally impose segregation of modified and non-modified crops but the provision will significantly help those countries (particularly in Europe) facing consumer pressure for transparency and clarity over the content of LMOs in shipments and products. Thus, consumer labelling would remain a domestic matter. While agreement on most of Article 18 had been easily obtained, this last issue regarding requirements for documentation accompanying LMO commodities became, somewhat surprisingly, the hold-out issue in the final hours of the negotiations. The matter was ultimately resolved through eleventh-hour bilateral consultations between the EU and the Miami Group – the latter characterising the documentation issue as a make or break point owing to its uncertain consequences in terms of production costs and segregation of crops.[10]

Status of non-Parties (Article 24)

Another sticking point for the Miami Group concerned the status of agreements with non-Parties (which are now only subject to the provisions in Article 24).

Transboundary movements between Parties and non-Parties must be consistent with the objective of the Protocol, but no further standards are imposed on agreements that may be entered into between Parties and non-Parties regarding such transboundary movements. The compromise reached accommodates the US's concern that it would be unable to negotiate the terms of any bilateral agreements with Parties to the Protocol. However, since it encourages the US to conclude agreements regarding the transboundary movements of LMOs consistent with the objective of the Protocol, it may be seen as an improvement over the current situation in which US exports of LMOs are not subject to any framework – bilateral or otherwise.

In order to permit agreement in Montreal, a range of significant questions was left for future elaboration by the Parties. These include the ways in which the Biosafety Clearing-House will operate, capacity-building for developing countries (Article 22) and, perhaps most difficult, the question of liability and redress for damage resulting from the transboundary movement of LMOs (Article 27). The future resolution of these issues will ensure substantive post-agreement negotiations following the Protocol's opening for signature at COP-5 in Nairobi in May 2000.

Conclusion

This chapter began with a consideration of the possible ways in which the new biotechnology might threaten the natural environment. It is arguable that the Biodversity Convention focus upon biosafety, as a primary area in which to develop a protocol was essentially misplaced – both environmentally and politically. From a conservation viewpoint, there are certainly more clear and present dangers to biodiversity. From a political perspective, to invest so much in a Protocol where major trade interests were engaged was certainly a high-risk strategy.

At the end of February 1999, following the acrimonious collapse of the negotiations in Cartegena, it appeared that the fears of sceptics were well-grounded. The question of the extent to which the Protocol should be a trade facilitation or an environmental protection agreement had not been answered and there was a very real possibility that the whole matter would be turned over to the WTO. Eleven months later it did prove possible to resolve the immediate issues and to agree the Protocol, even if the underlying trade/environment tensions remained. What emerged was a more precisely focused regime which will allow governments to exercise quite stringent precautionary controls over imports of LMOs (such as seeds or live fish) designed to be directly released into the environment. The key compromise involved the exemption, from AIA procedures, of other genetically engineered commodities, such as foodstuffs and pharmaceuticals, in which there was already huge commercial interest and very substantial international trade. Pharmaceuticals are not to be controlled at all while commodities for 'food, feed or processing' are subject to an alternative and much less restrictive procedures.

Whilst the Protocol contains certain 'constructive ambiguities' it may come to represent a major step forward in reconciling environmental, trade and development concerns with respect to the use of modern biotechnology. By giving legal recognition to the mutually supportive nature of trade and environment agreements, the protocol may help to better integrate environmental considerations into trade policy. How influential it proves to be within the WTO remains to be seen, however.

The final agreement at Montreal directs attention to the way in which the issue was defined and carried forward and to the political changes that allowed the differences, so evident at Cartagena, to be accommodated. The CBD itself was created amidst the politics of the wider UNCED process and was widely regarded as something of an achievement by the South, and as compensating for the dominance exercised by the North in the creation of the FCCC and other regimes. Subsequently, as negotiations clarified the outlines of a protocol, the unity of the developing countries that had characterised the negotiation of the CBD itself began to erode as national interests in biotechnology development and trade came into sharper focus (McGraw, 2000). The industrialised countries also took up varying positions, resulting in an important split within the OECD. The EU (notwithstanding major differences between its Member States) tended to move towards a more sceptical attitude towards the benefits and safety of biotechnology, and in any event defended its own precautionary procedures for LMOs. The US and the Miami Group maintained the view that anything more than a limited co-ordination of existing national regulations amounts to a restriction of trade based upon official certainty that the calls for a comprehensive international system to protect the environment from the unspecified dangers of LMOs transferred across boundaries is little more than unscientific alarmism, cloaking a number of ulterior motives.

Eleventh hour moves by the EU at Cartagena failed to broker a compromise, even though significant concessions had been made by the South on questions such as the inclusion of 'products thereof'. What had changed by January 2000? A number of factors are worthy of consideration. First, the broader context of the final year of negotiation was one of rising public alarm at the possible health and environmental implications of the new biotechnology. However inadequately grounded in science, they served to inhibit the actions of market-sensitive food manufacturers and, indeed, governments. Consumer resistance appeared to be making headway where the proponents of a moratorium or the regulation of 'products thereof' had failed. However, the US government and other Miami Group members remained fully aware of their extensive national interests in the new biotechnology and continued to react strongly against attempts by the EU to respond to the health and environmental fears of its citizens. Resort to the

WTO disputes procedure and more were threatened and, in the absence of an agreed Biosafety Protocol, it appeared as if the question would be fought out exclusively within the trade regime. The United States managed to convince the EU trade commissioner at the ill-fated Seattle Ministerial of the WTO to create a Working Group on biotechnology. Roundly condemned by EU environment ministers the proposal came to nothing as the whole attempt to set up the 'Millennium Round' foundered (*Guardian*, 3 December 1999). The failure to begin the incorporation of biosafety questions into the trade disciplines of the WTO may come to be seen as a turning point. Yet, there were other propitious factors as well.

The transparent and inclusive negotiating procedures adopted at Montreal contrasted, not only with the events at Seattle, but also with the preceding round of biosafety negotiations at Cartagena. The open access that all countries and sectors of civil society enjoyed, particularly in the Vienna Setting, promoted a collective sense of commitment towards both the process and the product. Thus, it was difficult for any government or interest group to stall the process, or to disown the result in the end. In this respect the Montreal negotiations may serve as a useful precedent.

Unlike those that created the CBD, the biosafety negotiations avoided polarisation along a strictly North–South axis. Rather than replicating the UN's traditional regional groupings, alliances were formed according to interest. The creation of the Compromise Group, itself accommodating various positions, was particularly constructive in this regard. One delegate described the group as an 'international lab' in which various proposals could be tested for broader agreement.

Finally, the conclusion of the Biosafety Protocol was crucial for the future credibility of the CBD, with its headquarters in Montreal. This fact was not lost on the Canadian government, which had to reconcile its dual role as hard-line spokesperson for the Miami Group and congenial host. At the same time the Clinton administration, aware of the rising popular concern about GM technology, was able to agree to a deal which was far less threatening to US commercial biotechnology interests, in terms of environmentally justified protectionism in its overseas markets, than that which had previously been on offer.

With evident relief, Klaus Toepfer, UNEP Executive Director, declared that 'the agreement goes a long way towards meeting the environmental concerns of the international community' (CBD, Press Release, 29 January 2000). The narrowness of the provisions of the Protocol make this a, perhaps understandable, exaggeration. How far the Protocol will go in meeting its own limited objectives will only be revealed when the regime for Biosafety becomes an operational reality and, perhaps most significant of all, when it is tested in relation to the powerful disciplines of the WTO trade regime.

Notes

1 As early as 1975, the IUCN Commission on Environmental Law (CEL) argued that the existing system of site- and species-specific legal instruments on biodiversity did not protect the natural environment as a whole. In 1985, the CEL drafted a global biodiversity convention which was circulated to governments as well as other NGOs and IGOS for commentary.

2 Several state and non-state actors have argued that making biosafety the focus of the first protocol under the CBD is not only misguided, but detrimental to other more pressing and proven threats to biodiversity.

3 Of course, genetic material can be found *ex situ* (mainly in the gene banks of the North). Access to *ex situ* collections is a major point of contention in ongoing negotiations to bring the FAO's 1973 International Undertaking on Plant Genetic Resources, which originally operated according to the CHM principle, into harmony with the CBD.

4 The fact that it was pushed forward at all as part of the Rio process was not unconnected with the affront felt by UNEP an its Executive Director Mostafa Tolba, when the climate change negotiations were taken over by the UN General Assembly (Brenton, 1994: 200; Hopgood, 1998: 170). Tolba was also emboldened by the success of the ozone negotiations which he had overseen and was determined that the CBD should not fail under his leadership (McGraw, 2000).

5 The CBD negotiations provided developing countries (who possess a preponderance of the biological assets under negotiation) with greater bargaining power than oher global issue areas – whether in trade or environment (McGraw, 2000).

6 The CBD was duly presented to the Senate for 'advice and consent' along with an extensive textual exegesis pointing out that the US had little to fear from a convention with which it was already in compliance. The Foreign Relations Committee 'reported out' favourably in 1994 but the Senate suspended consideration of ratification (USIS, Biodiversity and the Convention on 'Biodiversity', 5 January 1999).

7 Other groups which emerged from Cartagena include the Central and Eastern European countries (CEE) and the Compromise Group (comprised of Japan, Mexico, New Zealand, Norway, South Korea, Singapore and Switzerland).

8 The Cartagena text had contained an all-out savings or subordination clause (reversing the *'later in time'* rule set out in Article 30(3) Vienna Convention on the Law of Treaties). This proposal had been rejected by the EU and the Like-Minded Group on the grounds that it would have subordinated the Protocol to other international institutions.

9 Unlike the FCCC, the CBD has not created an additional subsidiary body on implementation – although some argue that such a body may well be needed as national reports and future protocols proliferate.

10 Rather than accepting the EU's proposal to delay implementation of the requirement to specify the identity of the LMOs no later than two years after the Protocol's entry into force, the Miami Group proposed that the COP itself decide 'on the detailed requirements' themselves within the same two-year time frame. In exchange for the EU agreement on this deferral, the Miami Group accepted a formulation stipulating that documentation for shipment of LMO commodities should state that they 'may contain' LMOs.

9

International trade conflicts over agricultural biotechnology

Robert Falkner

This chapter[1] covers several aspects of the international politics of biotechnology which are addressed in more detail elsewhere in this volume. It deals with the trade conflicts that have arisen in connection with advances in agricultural biotechnology and their introduction to global food production during the 1990s. Its main focus is on the complex nature of these conflicts, owing to the role played by a variety of domestic interest groups and the clash between commercial interests and societal values. This focus serves to remind us of the importance of domestic actors and processes in determining trade policy at the national level and the outcomes in international politics.

Introduction

The use of agricultural biotechnology in food production has become the focus of a new international trade conflict. In this conflict, the major exporters of genetically modified (GM) crops and food products are pitted against those countries that wish to impose stricter controls on GM imports for reasons of food safety and environmental concerns. Most prominently, the United States as the world's largest producer of GM crops has recently clashed with the European Union over its *de facto* moratorium on new approvals and imports of genetically modified organisms (GMOs) as well as its GM food labelling policy.

The commercial significance of this conflict can easily be seen from the present and estimated future size of the agribiotech market. GM, or transgenic, crops have been adopted by farmers world-wide at a rate that is unprecedented by agricultural industry standards. Between 1996 and 1999, twelve countries have contributed to an increase in the global area planted with GM crops by 23.5 times, from 1.7 million to 39.9 million hectares (James, 1999: i). At the same time, the global market for GM crop products has grown from a mere $75 million in 1995 to reach an estimated $2.1 to $2.3 billion in 1999, and is projected to climb to $8 billion in 2005 and $25 billion in 2010 (James, 1999: iv).

The political fallout of this dispute is equally significant. Echoing recent transatlantic trade disputes over the EU's banana import regime and its ban on the import of hormone-treated beef, the conflict over GM trade has further strained the transatlantic relationship at a crucial time in the development of the international trade system. The collapse of the World Trade Organization's (WTO) ministerial conference in Seattle at the end of 1999, as well as the failure so far to introduce a formal environmental mandate to the trade regime, only serve to highlight the importance of a working transatlantic relationship in the area of trade politics.

Given the political and economic clout of the US and the EU, it is not surprising to find their diplomatic rift at the centre of the international politics of biotechnology regulation, be it in the context of the negotiations on the Biosafety Protocol to the Convention on Biological Diversity (CBD) or the WTO talks in Seattle. However, the international conflict formation in the area of agricultural biotechnology is more complex than the notion of a 'conventional' trade conflict between Washington and Brussels would suggest.

The complexity of the GM trade conflict reflects the impact of 'new' policy issues on the trade agenda, as exemplified in the clash between international trade rules and global environmental concerns (Blackhurst and Anderson, 1992; Vogel, 1995; Brack, 1998). It is not simply diverging commercial interests within the corporate sector that are at stake in the GM conflict, but a more profound clash between commercial interests, on the one hand, and societal values, on the other. Accordingly, the international conflict formation includes, in addition to governmental trade officials and corporate actors, a wide and more diffuse range of national and transnational societal actors. This 'new' group of actors in trade politics, most notably environmental pressure groups, not only try to influence the official political agenda at the diplomatic level but also shape the socio-economic environment of international policy making through channels that exist in the global economy and global civil society (cf. Wapner, 1997).

Because of the complexity and the highly politicised nature of agricultural biotechnology, governments have seldom been in control of the international trade agenda on GMOs and have – more often than not – appeared to be only responding to developments in the global market and civil society. In order to understand the dynamics of the international trade conflicts over agricultural biotechnology, we therefore need to examine the interaction between trade diplomacy, corporate strategies and market structures, as well as societal developments.

The rapid process of biotechnological innovation and commercialisation by international biotechnology firms has created new policy dilemmas and demands for regulation that tend to over-stretch the regulatory capacities of many governments. New scientific and commercial developments are constantly changing the regulatory agenda, thus further accentuating the gap between technological innovation and political response (McGuire, 1999). Market dynamics and corporate strategies, and especially conflict among corporate actors (Falkner, 2000b), are important factors in the international politics of agricultural biotechnology.

Equally, societal developments and the activities of transnational non-governmental organisations (NGOs) have left their mark on the global GMO agenda. Consumer and environmental groups have succeeded in politicising agricultural biotechnology. They have lobbied for stricter safety standards, especially in Europe, and have put pressure on food producers and retailers to eliminate GM ingredients from their product range. In national and international campaigns aimed at biotechnology firms, NGOs have managed to undermine industry efforts to reassure the public of the safety of biotechnological innovation, thereby effectively changing the parameters of the global agribiotech market.

Seeds of discord: the growing transatlantic rift

The debate on genetic engineering in Britain has recently erupted into an emotionally charged dispute over the environmental and health aspects of the new transgenic products: genetically modified organisms and food products containing genetically modified ingredients. National newspapers have run front-page reports about the dangers of so-called 'Frankenstein Foods', and widespread consumer concern has caused a large number of food producers and retailers to drop GM ingredients from their menu. Not only environmental pressure groups, but also prestigious institutions such as the British Medical Association have called for a moratorium on commercial releases of GMOs into the environment (*Financial Times*, 1999c; Hileman, 1999).

While the UK government is adamant in its effort to reassure the public of the safety of GM products (*Guardian*, 1999b; *Financial Times*, 1999f), large parts of society appear to have lost confidence in the scientific community and regulatory authorities. Scepticism over GMOs in the UK reflects a wider crisis of confidence in the industrial production of food, and the role played by biotechnology in particular, which can be traced back to previous food scares revolving around BSE ('mad cow disease') or salmonella-infected eggs (*Guardian*, 1999a). But the situation in the UK is not unique; growing unease about the potential dangers of biotechnology can also be found in continental Europe (Süddeutsche Zeitung, 1999).

The public backlash against developments in biotechnology has caused governments in Europe to rethink their approach to regulating this new technology. During the 1980s, most governments in the EU were primarily concerned with promoting biotechnological research and the creation of corporate enterprises to capitalise on scientific innovation. Government officials have long regarded biotechnology as one of the key technologies of the future that could introduce important new growth dynamics into the European economy. Part of this effort was to re-examine, and where possible reduce, the regulatory burden that was seen to hold back the commercialisation of biotechnology in Europe. But public concern over the safety of GM food forced European governments to strengthen the regulatory regime in the 1990s.

This stands in sharp contrast to the situation in the United States, the leading centre of biotechnological research and production. Unlike Europe, the biotechnology industry in the United States has until recently experienced little public resistance to the introduction of new transgenic products. US consumers can choose from a variety of food products containing genetically modified potatoes, tomatoes, soya beans, maize and canola. Opinion polls suggest that the majority of consumers welcome the benefits that come with GM foods, and a significant number are simply unaware of the widespread presence of genetical engineering in food (*Financial Times*, 1999a).

The reasons for this transatlantic gulf in societal perceptions of biotechnology are manifold, and do not simply reflect differing levels of environmental concern. For one, food and certain aspects of food production play a bigger role in European society than in the United States. Furthermore, a number of food-related scandals such as 'mad cow disease' have undermined consumers' confidence in regulatory bodies in Europe. And owing to environmentalists' campaigns and labelling schemes, European consumers tend to be more informed about the presence of genetically modified food ingredients than their American counterparts.

These cultural and political differences between Europe and the United States are mirrored in the different approaches to biotechnology regulation. The European Union issued two directives in 1990 that concern the contained use of GMOs and their release into the environment, requiring environmental evaluation and step-by-step approval for the dissemination of GMOs (European Council, 1990). The EU also requires foods containing GM ingredients to be labelled as such. The EU regulations give extensive rights to member states to participate in the evaluation process but leave the final decision on, for example, the market introduction of GM products with the Commission. These procedures have been criticised by the biotech industry for producing a far too complex and unpredictable approval process that gives priority to political over scientific considerations. In contrast, the United States' more streamlined and de-politicised evaluation procedure has allowed industry to promote the rapid commercialisation of biotechnological research. By the end of 1998, there were about thirty transgenic plants on the US market.

These differences in regulatory practice have been cited as one reason for the global dominance of the US biotechnology industry and the relative weakness of European firms. Moreover, European biotechnology firms have invested heavily in the United States, thus further accentuating the US's leading position in this industrial sector. As a consequence, the American biotechnology sector has emerged as the world's biggest exporter of GMOs, while the European Union remains a GMO-importing region. It is estimated that in the United States 28.7 millions hectares of agricultural land have been planted with GM crops in 1999 (72 per cent of global area of GM crop production). In contrast, only Spain and France produce GM crops in commercial quantities though on a relatively small basis, each totalling less than 1 per cent of the global GM crop area (James, 1999: 6).

The European Union has now come under pressure from two sides to streamline its biotechnology rules. Its own industry, which has recently co-ordinated its lobbying activities and created the EU-wide industry association EUROPBIO, has mounted a concerted campaign to resist further restrictions on the European biotechnology sector. EUROPBIO has called for a clearer and more predictable set of rules regarding GMO research and commercialisation, to replace the existing EU directives. It successfully lobbied for European Parliament approval of the Commission's Directive on the Legal Protection of Biotechnological Inventions in 1997.

At the same time, the US government and US biotech firms are urging the European Commission to eliminate barriers to American GMO exports. American trade officials have repeatedly complained about the cumbersome and discriminatory regulatory process in the EU and its labelling regime for GM foods. Stuart Eizenstat, State Department Undersecretary for Economics, told a US Senate finance committee in March 1999 that the EU approval system for GMOs was 'non-transparent, unpredictable, not based on scientific principles and all too susceptible to political interference' (*Financial Times*, 1999d). Both the US government and US biotechnology firms accuse the EU and its member states of politically motivated obstructionism in the regulatory approval process which they claim has cost the US corn growers some $200 million of annual export sales in 1998 and 1999 (Larson, 1999).

But domestic pressure by environmentalists and the European Commission's desire to strengthen its role in maintaining food safety in Europe make it unlikely that the EU will simply give in to corporate lobbyists from both sides of the Atlantic. To be sure, the Commission is not united on this issue. Within the Commission there are advocates of a liberalisation of the EU's regulatory system who consider the regulatory framework agreed in 1990 as a flawed system which was pushed through in a single-minded effort by environmentalists without wide-scale consultation. Yet, as consumer and environmental groups step up their demands for a ban, or at minimum a moratorium, on the release of GMOs into the environment and their use in food products, the Commission is coming under increasing political pressure to further increase restrictions on the biotechnology industry. Recent decisions by the Commission underline these political pressures: a *de facto* moratorium on approvals of new GMO products within the European Union is unlikely to be repealed before 2002, and the Commission decided in January 2000 to require compulsory labelling of food products which contain more than 1 per cent GM ingredients (*ENDS Daily*, 2000).

International biosafety negotiations

Besides the contentious European regulatory framework, the conflict between the United States and the European Union has centred on the creation of an international biosafety treaty (see also Vogler and McGraw, chapter 8, this volume). Biosafety talks had been underway since 1996, largely unnoticed by the media and the general public. The safety of biotech trade became the first issue area to be regulated in the form of a Protocol to the 1992 Convention on Biological Diversity (CBD). Although a number of international organisations have been concerned with various aspects of international biosafety, among them UNEP, UNIDO, FAO and OECD, none of them managed to provide more than incomplete or voluntary codes of conduct.

The biosafety talks went into the final round in February 1999, against the backdrop of growing public concern over GM foods in Europe. The 138 negotiating parties that met in Cartagena, Colombia, were faced with a heavily bracketed draft treaty that contained numerous unresolved issues. Despite reaching some compromises, the parties failed to conclude negotiations and decided that the talks had to be suspended in light of insurmountable differences between the group of GMO exporting countries, on the one hand, and the EU and developing countries, on the other.

The conflict in Cartagena centred on, *inter alia*, the questions of how far the proposed Biosafety Protocol should interfere with the principle of free trade; which types of GMOs should be covered; and on what basis importing nations should be allowed to refuse GMO imports. The groups of GMO exporters, the so-called Miami Group, insisted that the Protocol be subordinate to other international agreements and obligations, that agricultural commodities be excluded from the scope of the treaty and that decisions on GMO imports need to be based on 'sound' scientific assessment.

In sharp contrast, the majority of developing countries, the so-called Like-Minded Group, argued that the Protocol be exempt from WTO disciplines, that all types of GM products be covered, and that precaution should guide the risk assessment process. The EU supported many of the demands of the Like-Minded Group and proposed a compromise package. This was, however, rejected by the US-led coalition of GMO exporters, and the suspension, rather than break-down, of the talks was agreed as a face-saving measure.

Despite mutual recriminations after Cartagena, the parties continued to discuss the scope for a compromise and met three times before the Extraordinary Conference of Parties resumed in Montreal on 24 January 2000. By that time, significant progress had been made, and the Montreal negotiations succeeded in removing the last hurdles for the adoption of the Cartagena

Protocol on Biosafety (see Falkner, 2000a). The outcome of the Montreal talks represents a significant achievement in bridging the North American and European positions on biotech trade. Agribiotechnological commodities were included in the Protocol, although less strict obligations for notification and labelling apply to them. And, crucially, the EU succeeded in having the precautionary principle acknowledged in the context of the treaty's risk assessment procedure.

However, the transatlantic biotech dispute is far from resolved. The Protocol's provisions on its relationship with other international agreements leave room for interpretation and do not rule out a potential referral of disputes over biotech trade restrictions to the WTO. Not being a party to the CBD due to strong opposition in Congress, the US government is unlikely to achieve ratification of the CBD and the Cartagena Protocol in the near future. The treaty therefore does not create binding obligations for the US, which is likely to continue to resort to bilateral efforts to seek market access for its GM exports.

The North–South dimension

While the Cartagena Protocol has given the EU's regulatory system greater international legitimacy, its main function will be to support developing countries in their efforts to enforce biosafety controls on GMO imports. The negotiations on the Biosafety Protocol laid open a different conflict formation in the international debate on agricultural biotechnology that is likely to gain in importance in the future. In the run-up to the Cartagena and Montreal Conferences, many developing countries expressed their grave concerns over the way in which commercial interests of industrial countries dominated the discussion of international biosafety rules at the expense of developing countries' interests. Most developing countries supported the adoption of the Biosafety Protocol in the hope that it would strengthen their regulatory powers *vis-à-vis* transnational biotechnology firms. However, the conduct of negotiations underlined their fears that the Protocol would focus more on facilitating trade in biotechnological products than on establishing stringent and far-reaching safety norms.

The negotiations also demonstrated the limited diplomatic clout of developing countries in influencing the outcome. For most of the biosafety talks, the developing world failed to present a united front, and ultimately depended on the EU's support in trying to extract concessions from the group of GMO exporting countries. Political divisions that reflect the growing diversity of interests in the South initially undermined the negotiating position of the developing world. At the start of the biosafety talks, the group of developing countries failed to speak with one voice. A small number of states that had moved into GM crop production during the second half of the 1990s, most prominently Argentina, began to side with the northern GM crop-exporting nations. The differences within the group of southern states were eventually resolved when in early 1999 Argentina, Chile and Uruguay joined the Miami Group of GM exporters and the G-77 and China formed the 'Like-Minded Group', giving a united voice to the majority of developing countries (Egziabher, 1999).[2]

In order to understand the developing countries' divergent positions on the Biosafety Protocol, and the lack of a common political strategy, it is necessary to reflect on the opportunities and problems that modern agricultural biotechnology creates for the developing world.

A few countries, such as Argentina, Chile and Uruguay, have embraced the new technology and developed a strong interest in promoting GM crops as part of their agricultural export strategy. They lend diplomatic support to the United States and Canada in their effort to limit the interference in international trade by any international biosafety standard. Others, such as Brazil, initially welcomed agricultural biotechnology but have more recently reconsidered their crop export strategy. A Brazilian court ruling in June 1999 forced Monsanto to stop selling its genetically modified 'Roundup' soya and to carry out an environmental impact study. Demand from Europe for non-GM soya has since caused some Brazilian farmers to reject the GM crops in fear that they would jeopardise an important export market (*Financial Times*, 1999e). Internationally, Brazil has taken a stronger position in support of a comprehensive Biosafety Protocol, thus bringing it directly in conflict with its neighbour, Argentina.

Most developing countries, however, are not in a position to develop their own biotechnology research capacity or attract substantial international biotechnology investment. The main problems that these countries face are those of importers of GM crops or test grounds for the release of GMOs into the environment. They may wish to promote biotechnology as part of their agricultural strategy, but will fear about the safety of GM imports and their use, especially given the widespread lack of scientific and regulatory expertise in this field.

The prime benefit of agricultural biotechnology to developing countries lies in the creation of a more productive agricultural sector. By raising productivity, GM crops can potentially contribute to a country's overall economic development as well as help to alleviate food shortages and malnutrition. GM crops can reduce the costs of production and increase agricultural yield by making crops more resistant to diseases. Genetic engineering can also increase the nutritional value of crops, for example by enriching them with minerals and vitamins (Nuffield Council on Bioethics, 1999).

The realisation of these benefits depends on many factors, including the availability of biotechnology programmes that promote the adoption of the new technology in agriculture. Currently, however, biotechnology research in the developing world is taking place in only a few countries, such as

China, Brazil, Egypt, India and South Africa. Most developing countries lack such programmes or simply do not have the economic, scientific or institutional resources to develop a comprehensive national biotechnology strategy.

Developing countries are also concerned that the benefits of investing in biotechnological innovation could be offset by the potential risks of the new technology. Among those risks, safety aspects of agricultural biotechnology are of particular importance in the developing world, where scientific, legal and regulatory capacities are on the whole inadequately developed. In a rapidly changing research environment, many poor countries would be unable to exercise effective control over transnational biotechnology firms that introduce GM products or applications, unless they receive international support to build up regulatory capacity. Moreover, representatives of developing countries have expressed the fear that they might become 'dumping grounds' for biotechnological developments which do not pass the more stringent safety standards in industrial countries (Sasson, 1993: 29–30). It is for these reasons that developing countries have invested great hopes in the Biosafety Protocol.

But concern for biosafety alone does not adequately describe the interests of developing countries. Issues relating to national sovereignty, economic competition and the relationship between private and public actors also play an important role.

For one, the dominance of corporate actors from industrial countries in biotechnological research, commercialisation and trade has given rise to fears that developing countries would loose control over their food production systems if they allowed GM crops to permeate southern agriculture. This fear has been further heightened by the fact that the WTO's intellectual property rights (IPR) system, as represented by the TRIPS agreement, forces developing countries to import northern IPR norms and rules into their national legal systems (see Williams, chapter 5, this volume). This gives transnational biotechnology firms a powerful instrument in their attempt to patent genetic resources that have hitherto been used freely by farmers in the developing world. It also raises the spectre of increased dependency on northern firms if southern agricultural producers adopt GM crop varieties (Shiva and Moser, 1995).

Agriculturalists in the South also fear that traditional farming systems would be eroded by the new technology. The adoption of a narrow genetic base of a few internationally traded GM crops could contribute to the loss of biological diversity, local knowledge and sustainable agricultural systems.

Furthermore, the changing role of the public and private sectors in agricultural biotechnology research funding has given cause for concern in developing countries that their special needs are not adequately reflected in biotechnological innovation. While the development of 'Green Revolution' technologies was funded primarily by public institutions and philanthropic foundations, the trend in agricultural biotechnology research has been towards a larger role of corporate institutions (Brenner, 1997: 8). Given the high costs and commercial risks involved in biotechnological research, the research done by northern firms focuses on the agricultural sectors of industrial countries, where they expect the highest rate of return on their investment. One result of this imbalance between private and public sectors in biotechnological research has been that the genetic modification of crops, in order to improve the nutritional value or the resistance of crops in developing countries, has been relatively under-researched.

The changing dynamics of the GM trade conflict

The successful completion of the biosafety talks in Montreal has given a boost to developing countries' efforts to create effective biosafety regulation. In the context of the transatlantic relationship, the agreement on the Protocol has helped to reduce the level of tension, although it failed to completely remove the underlying cause of conflict. The US continues to view European biotech regulations as unjustified trade interference, and will try to make sure that its viewpoint is reflected in the implementation and further evolution of the international biosafety regime. The EU, on the other hand, feels its approach to biotech regulation has been vindicated by the outcome in Montreal. In forthcoming negotiations to specify labelling requirements for GM commodities under the Cartagena Protocol, the EU will therefore try to strengthen the Protocol's provisions. The decision in October 1999 by the EU's European Standing Committee on Foodstuffs to introduce compulsory labelling for foods where at least one ingredient contains more than 1 per cent of genetically-modified material ingredients underlines the determination of the EU to stand firm on this issue.

Given these persistent political differences and the fact that the US is not a party to the Biosafety Protocol, GM trade will continue to give rise to tensions between the US and the EU. To understand the changing dynamics of the GM trade issue in transatlantic relations, it is important to emphasise that, unlike in many other cases, the trade restrictions at the centre of the dispute are not simply protectionist in nature. Although the US has initially viewed European objections to unfettered international trade in GMOs as a case of scientifically unfounded protectionism, the European Union's regulations are responding primarily to genuine consumer concerns about the safety of GM products.

At the centre of the conflict are differences in societal values with regard to the risks and benefits of biotechnology. Americans highlight the wide-ranging gains from genetic engineering, such as higher efficiency and disease resistance of GM crops, and consider the risks to be scientifically uncertain. Europeans, in contrast, tend to question the need for more intensive forms of industrial agriculture and want to see the precautionary principle applied. Even if a diplomatic compromise between the EU and the US can be found, it may not satisfy consumers in Europe.

Equally, domestic pressure groups in the US are unlikely to give in too easily to European demands for environmentally inspired restrictions on GM trade. The continuing closure of the European market to GM imports would not only deprive the US biotechnology industry of the most important export market but would also set a dangerous precedent for other potential markets, in Asia, Africa and Latin America. Measures to restrict GM imports have already been taken by Thailand (Reuters 1999a), and Brazil, South America's biggest market, has recently adopted a more cautious approach on GM crops. Faced with the threat of reduced growth prospects, or even a gradual closure, of export markets for American GM products, the US biotechnology industry will employ all its power to lobby against GM trade restrictions. Furthermore, faced with falling prices and the collapse of Asian markets, the agricultural sector in the US has put pressure on the Clinton Administration to resist protectionist tendencies around the world and has already successfully lobbied for a tough US stance on bananas and beef (Granville, 1999).

The contours of the evolving GM trade dispute between America and Europe depend to a large extent on developments in the political economy of agricultural biotechnology. Changes in the balance between divergent corporate interests play a crucial role in the evolution of trade policy as well as foreign environmental policy (Falkner, 2000b). The biotechnology sector in the USA is likely to exert strong influence over Washington's trade policy if it can present a strong case that is backed by a relatively united corporate front, as has been the case in the recent past. Cracks in the corporate front, however, and the emergence of powerful countervailing interests, would alter the domestic basis of the US position on GM trade.

Indications of such changes have recently emerged, and it can be argued that they have contributed to the acceptance of the Cartagena Protocol by the US government. For one, the American environmental movement has begun to make agricultural biotechnology a campaigning issue after having almost completely ignored it during the 1990s. Partly inspired by the success of their European organisations, Greenpeace and Friends of the Earth in the USA have begun to raise awareness among American consumers about the potential health and environmental risks involved in genetically modified food production. An indication of the impact of these campaigns is the admission of the Food and Drug Administration (FDA) in October 1999 that it needed to consult more widely with the public on its approval policy for GM foods. Faced with growing consumer concerns, the FDA began to hold a number of consultative meetings across the United States in which it aimed to inform the public about the agency's food safety policy (*Christian Science Monitor*, 1999a).

The environmentalists' campaigns are likely to increase in intensity over the next few years. Whether the American public will ever turn against GM foods to the extent that European societies have done is uncertain. But the

American biotechnology industry now knows that the period when it faced virtually no public resistance to its developments in the field of agriculture is over. Growing consumer resistance against GM crops and foods has already translated into market signals that have sent shock waves through the biotechnology sector.

In an assessment of the social and political risks involved in agricultural biotechnology, a recent Deutsche Bank investment report came to the conclusion that 'the term GMO has become a liability', and that 'GMOs, once perceived as the driver of the bull case for this sector, will now be perceived as a pariah'. The industry analysts saw a separation of the grain market developing, with GM corn and soybeans selling at a discount to non-GM crops, and accordingly reduced their investment rating of GM seed companies (Ramey, Wimmer and Rocker, 1999). Against the background of growing public controversy in the United States and future profit warnings for agricultural biotechnology (*Financial Times*, 1999b), the publication of the Deutsche Bank report caused an immediate fall in the stock market value of US biotechnology companies. It sent unequivocal signals to investors that agricultural biotechnology no longer held the promise of exceptional future growth rates.

The growing unease among biotechnology investors helped to speed up the consolidation and concentration process in the sector by strengthening the pressures for corporate restructuring. Monsanto and Pharmacia & Upjohn, two leading biotechnology firms, agreed in December 1999 to merge and to have their joined agricultural biotechnology business independently listed in the stock market in order not to burden their more profitable pharmaceutical biotechnology business with the risks of the GM food controversy. The decision to merge was widely interpreted as indication of the damage that Monsanto had suffered in its agri-biotechnology business, including an anti-trust lawsuit brought against the company and other biotechnology firms in late 1999. Other companies, such as the Swiss multinational Novartis, are also reported to be searching for a merger partner (*The Daily Telegraph*, 1999, *Financial Times*, 2000).

Just as the biotechnology industry is undergoing a process of restructuring, developments in the US agricultural sector further raise the spectre of a fragmentation of the business community on key issues in the GM debate. In the past, the farming lobby stood more or less united behind the biotechnology sector in its attempt to gain access to the European market and block international efforts to create a comprehensive Biosafety Protocol. Having adopted a variety of GM crops at a rapidly growing rate, American farmers had a clear commercial interest in supporting the biotechnology industry in its opposition against international trade restrictions as well as GM labelling schemes. The latter has been resisted on grounds that it is scientifically unwarranted, but also practically impossible to achieve, as US farmers have usually not separated GM crops from non-GM crops.

But with the hostile reaction of European consumers to GM crop and food imports and growing concerns among US consumers, American farmers had to rethink their strategy. Despite considerable costs to grain handling and food processing firms, a trend has started in the US towards segregation of GM and non-GM produce. This would remove an important obstacle to the introduction of GM labelling in the US. Moreover, the take up of GM crop production in the United States is predicted to slow down over the next few years, partly as a sign of growing concern among farmers over consumer reaction and the closure of international markets (*Christian Science Monitor*, 1999b; *Washington Post*, 1999, *New York Times*, 1999).

At the same time, a coalition of 60 US consumer groups has launched a campaign for GM labelling, and a cross-party coalition of US congressmen introduced a GM labelling bill in November 1999 (*The Guardian*, 1999c). The market changes in response to global civil society action may well turn out to be a turning point in the US policy on labelling, both domestically and internationally, and have already contributed to a gradual softening of the US's stance on international biosafety standards. It is too early to judge the effect of these developments on US foreign policy. They nevertheless illustrate the growing complexity of interest formation in the GM trade conflict and the importance of societal and market developments in bringing about political change.

Outlook: biotechnology on the WTO agenda?

While changes in the US market for GM products are underway, Washington continues to urge other nations to open their markets for American GM exports. Having unsuccessfully used unilateral pressure on the European Union, the US government has recently escalated the conflict by pushing for inclusion of biotechnology issues in the agenda of the World Trade Organisation before agreeing on the Biosafety Protocol in January 2000. The US has so far abstained from bringing a case against the EU's policy on GM imports to the WTO. Having the WTO consider international rules and disciplines regarding trade in biotechnological products could potentially shift the balance in the international GM trade dispute in favour of the US. The question of whether, and how, the WTO would deal with trade aspects of agricultural biotechnology is therefore central to the future evolution of the US–EU and North–South conflicts over GMOs.

What, then, are the prospects of the biosafety issue getting on to the WTO's agenda? In the run up to the WTO conference in Seattle, two proposals were made to formally introduce biotechnology to the WTO: Canada and Japan suggested the establishment of a WTO working group on GM food, while the United States submitted its own plan for discussions within the WTO to create additional disciplines for trade in biotechnological products. The objective of the US proposal is to ensure that 'trade in products of agricultural biotechnology is based on transparent, predictable and timely processes' (Revised draft of WTO Ministerial Text, 19 October 1999, section on agriculture, paragraph 29) – a barely concealed effort to address the problems arising from the EU's regulatory system.

The EU has so far avoided a clear-cut response to these initiatives and tried not to jeopardise the Montreal negotiations on the Biosafety Protocol by opening up WTO discussions on biotechnology. However, divisions within the European Commission surfaced during the Seattle conference when EU trade commissioner Pascal Lamy reached a bilateral agreement with the US on the inclusion of biotechnology in the WTO agenda, only to be rebuked by the EU's General Affairs Council shortly afterwards for having broken with, and undermined, the existing policy consensus in the EU.

The successful outcome of the Montreal negotiations on the Biosafety Protocol has, for the moment, put on hold the proposals to deal with biotechnology trade issues under the WTO umbrella. But the US, the world's largest GMO exporter, is unlikely to become a party to the Cartagena Protocol in the near future. There remains the potential threat of legal action under the WTO's trade rules by a GMO exporting country that feels that GMO import restrictions or labelling schemes lack proper scientific justification and are therefore discriminatory.

The current proposals for a WTO working group or additional disciplines are not sufficiently detailed to suggest in what way the WTO's involvement would change the nature of the GM conflict. In the run up to the Montreal biosafety talks, observers expressed concern that a transfer of the biosafety issue from the Convention on Biological Diversity to the WTO would significantly alter the political agenda (Stilwell, 1999). Especially developing countries feared that the WTO's emphasis would be more on facilitating biotechnology trade rather than the creation of biosafety standards; that the burden of scientific evidence in regulating GM trade would shift towards the importing country; and that new WTO disciplines could be used to counterbalance the regulatory provisions of the Biosafety Protocol.

Whether or not the WTO formally considers the regulation of trade in biotechnological products, the potential clash between trade rules and environmental regulation will play an important role in the future evolution of the biosafety regime. The adoption of the Cartagena Protocol has, for the moment, reduced the chances of a WTO biotech working group being set up, but the issue may resurface in future WTO trade talks. The fudged compromise on the relationship between the biosafety regime and the international trade order may have helped to salvage the biosafety talks in Montreal, but the issue of how to reconcile trade interests and environmental concerns in the area of biotechnology is bound to re-emerge on the international agenda. Although trade representatives are generally wary of environmentally dressed-up forms of protectionism, concerns over the safety of GM products and the claims of states to the right to decide about GM imports need to be taken seriously. The biosafety controversy is the clearest indication that societal values with regard to human health and the environment will increasingly compete with commercial interests on the trade agenda.

Notes

1 This chapter builds on the discussion of trade and agricultural biotechnology in Falkner (1999).
2 Most of the world's biotechnological research is concentrated in the industrial world, as is the production of GM crops: It was estimated that in 1999, 82 per cent of the global area of GM crop production was in industrial countries, and 18 per cent in developing countries (James, 1999: iv). Of all developing countries, only Argentina produces a sizeable portion of GM crops (17 per cent of global area of GM crop production area), with China and South Africa each accounting for less than 1 per cent of the global GM crop production area.

Part III

Biological warfare:
prospects and responses

This part provides marked contrast with the last part – drawing us into security concerns given the quite different potential of the same technology to underpin certain categories of biological weapons. At the core of the chapters that follow is the practice of deploying biological materials as military weapons against biological organisms – human or otherwise. In that biotechnology is concerned with manipulating biological materials, there are problems of defining its limits (Manning, chapter 2, this volume). Old and new practices are involved in biotechnology: brewing as an ancient technique now benefits from genetically modified yeasts; agriculture is being offered GM seeds; the treatment of illnesses is being updated by better understanding of genetic processes, the development of GM-based diagnostic tests, GM-derived medicines and, gradually, direct genetic treatments. Old and new dimensions of biotechnology have come together.

The same is true in the development of biological weaponry. The use of weapons derived from biological sources, like much of biotechnology, is not new. Spreading disease by flinging carcasses over the defended walls of castles or towns may have been replaced by more sophistication in biological weapons, extending to the application of genetic manipulation techniques in designing novel weapons. But the basic principle is the use of biologically based materials. Thus, Dando (chapter 10) opens the part by reviewing the 'classical agents' that can be used in the deliberate spreading of disease to cause illness and death. This review illustrates both the scale of research programmes in various countries and the wide range of potential agents at the disposal of those so disposed to seek such weapons to target humans, animals or plants. He finishes his overview with a look towards a future where, not least, genetic manipulation, as an advance in scientific sophistication, brings forth new classes of weapons.

Dando's review is complemented by Whitby's (chapter 11) survey of anti-crop biological warfare and the implications for the Biological and Toxic

Weapons Convention (BTWC). His chapter reminds us that biological weapons need not be directed against humans and animals but against crops and that, like other weapons types, there is a history showing considerable effort on behalf of some states to research and develop plant pathogen-based weapons. In passing he identifies some of the difficulties in delivering biological weapons. Recognition of the threat of anti-crop weapons is also shown through an examination of the BTWC review process and associated working party documents. In discussing potential responses to the threat, Whitby notes the importance of surveillance supported by international agreements, including the Biosafety Protocol (see Vogler and McGraw, chapter 8, this volume). Less is said about the direct influence of genetic manipulation techniques, although the importance of potential applications is recognised – a potential founded upon an older tradition and expertise.

Whatever the biological weapon, whatever the type of target, and whoever the possessor, a question arises over the collective international response to the problem. Generally accepted as a class of weapon in need of control, in the sense of non-proliferation, and to be subsumed within an appropriate international regime, much attention has come to rest with the BTWC. Littlewood (chapter 12) centres his analysis on this convention while recognising other responses. Indeed, he begins with the recognition that the development of biological weapons by states has in some cases been in response to the fear that other states were developing them – namely a motivation to develop a deterrent posture (a theme developed by Spear, chapter 13). In this as other areas of technology progress outstrips regulation. Littlewood adds a further slice of history regarding this class of weapon to that of the previous two chapters, reinforcing, once again, the importance of older technology that in turn may be enhanced by developments from the *new* biotechnology. In particular the BTWC was agreed before such developments came on the scene. It was also flawed from the start because of the absence of any verification process. Largely because of this and the ability of technology to outpace regulatory efforts, Littlewood is pessimistic about the long-term success of the BTWC process.

In the final chapter (13) of this part, Spear is not convinced that the BTWC can cope with the overall threat of biological weapons, in particular because of the underlying assumption of the Convention that states are the source of the problem. With non-state actors potentially able to acquire and deliver a biological weapon in a world that is increasingly transnational, the convention, with its limitations already identified by Littlewood, may simply be overwhelmed. Setting the problem in a wider security perspective, Spear proceeds to analyse alternative responses to the problem which include deterrence, detection and defensive measures.

Taken together the four chapters provide a contrast to the international political economy concerns of the last section. In some respects this arises from the generic nature of the technology, different trajectories of development, some of which were established before the new biotechnology arrived, and the duel-use character of particular technological advancements. The security concerns of this section may bring terrifying prospects to the fore, if we contemplate the use of biological weapons as a weapon of mass destruction. While the concerns that arise in IPE terms may be less immediate but nonetheless able to invoke very different visions of the future.

10

Biological warfare:
the 'classical' agents

Malcolm Dando

Introduction

During the course of evolution a variety of pathogens have become capable of causing disease in humans and important animal stocks and food crops. In the disrupted social conditions of large-scale warfare these pathogens have often caused huge numbers of deaths and agricultural losses. This, however, has not been the directly intended result of human action.

Biological warfare, on the contrary, is the *deliberate use of disease* to cause illness and death. There appear to be examples of such actions in historical record prior to the end of the last century, but it was only then when scientists such as Pasteur and Koch demonstrated that specific micro-organisms caused specific diseases in humans, animals and plants that modern scientifically based biological warfare became possible.

In such circumstances it is not surprising that we have seen a series of offensive biological warfare programmes carried out by the military forces of a number of important states during this century (Dando, 1999: 43–62). This process began with efforts by both sides during the First World War to damage the valuable draft animal stocks of the other by the use of biological agents. It continued in the inter-war years, particularly in the massive, terrible, Japanese offensive biological warfare programme in China, which culminated in a number of attempts (field trials) to use biological agents against human beings on a very large scale.

It was only, however, when the British, during the Second World War, having produced a crude anti-animal weapon based on anthrax, turned their attention to a scientific study of how to use micro-organisms most effectively and discovered that attack through the lungs was most effective, that the real potential threat of anti-personnel and anti-animal biological warfare was realised. The British then joined forces with the USA and Canada and the USA continued its massive offensive biological warfare programme through to the late 1960s. This led to the weaponisation of a number of 'classical' anti-personnel and anti-plant agents.

Following the closure of the US offensive programme, and the agreement of the 1972 Biological and Toxin Weapons Convention, the former Soviet Union dramatically increased its secret, and unlawful, offensive programme. This clearly used the growing knowledge of very dangerous viral agents and the beginnings of the new revolution in biotechnology (genetic engineering) (Nathanson, Darvell and Dando, 1999). Recently it has also been shown convincingly that Iraq had an offensive biological warfare programme prior to the 1991 Gulf War and so did white-ruled South Africa before it collapsed. Whilst the openly available information is not clear cut it is certainly believed by many informed people that there are a number – perhaps up to twenty – other countries which today have some form of offensive biological warfare programme.

This then, whilst still a not widely understood form of warfare, is an important threat that needs to be minimised by all possible means (Pearson, 1993). It is also a threat which is likely to grow as capabilities in the new biotechnology, for good medical, agricultural and industrial reasons, spreads around the world.

Anti-plant biological warfare is dealt with in the next chapter (Whitby). Here we concentrate on anti-personnel and to a lesser extent anti-animal biological warfare. The main intention is to describe the 'classical' biological agents and how their use could be envisaged. The chapter ends by briefly considering how this threat might develop in the next century if effective international arms control cannot be put quickly into place.

Pathogens

A number of complex multicellular organisms can cause disease in animals and humans. Examples are the tape worm parasites, and the causative agent of malaria. However, because of the need for rapid delivery and effect in military operations, considerations of biological warfare are confined to the use of micro-organisms: bacteria, viruses and some toxins. Fungal agents are predominantly used in anti-plant warfare.

Even with these restrictions there are a large number of pathogens that might be considered by the designers of biological weapons. Some examples of well-known pathogens are given below (Dando, 1994).

Cholera
This is a diarrhoeal disease caused by the bacterium *Vibrio cholera*. The bacterium gets into the water supply from an infected persons faeces and is then able to infect other people. It adheres to the intestinal wall and secretes a toxin which causes the severe diarrhoea. Victims lose huge amounts of fluid and many people die if not treated. Whilst cholera can be treated today by

modern methods it caused huge numbers of deaths in the last century and remains a threat to populations without adequate public health measures.

Measles

Measles is caused by a virus of the genus *Morbillivirus* of the family Paramyxoviridae. It only infects human beings. The virus is very contagious and is spread through the air or by direct contact. Infants can get passive immunity from maternal antibodies for six to nine months after birth and infection or vaccination confers lifelong immunity.

Obviously with such characteristics it is likely that measles will be an endemic disease which causes epidemic outbreaks every few years in an unvaccinated population. Mortality is highest amongst the very young and very old and can reach 5–10 per cent in poor conditions. At present there are some fifty million cases and 1.5 million deaths annually world-wide as a result of measles.

Foot-and-mouth disease

Foot-and-mouth disease (FMD) is a very highly contagious disease of cattle, pigs, sheep, goats, etc. It is caused by the FMD virus of the Aphthovirus genus of the Picomaviridae family. There are seven serotypes of the virus and no cross-immunity between the different types. The virus causes severe production losses and can cause 50 per cent deaths in young animals.

Major regions of the world are now free of the disease and extreme measures are taken to try to ensure that the virus does not return to such regions. The disease however is enzootic in South America, the Middle East, Asia and Africa and most wild cloven-hoofted animals, as well as some other animal species, are susceptible.

The virus is not only contagious, but can also travel on the air for long distances or remain alive for weeks in contaminated straw or manure. Infected animals rapidly acquire painful ulcers particularly around the mouth and feet areas. Fever, severe loss of appetite and cessation of milk production result from the infection. The US did not actually weaponise any such anti-animal agent but they could obviously be used to cause great economic damage to an agricultural industry in a covert attack today.

Agent selection

It would, of course, be possible for a small amount of a potent toxin to be used in an assassination attempt, or for a terrorist to attempt to contaminate food with a biological agent, but for a significant anti-personnel military strategic effect, or use as a weapon of mass destruction against civilians, biological agents have to be spread in the air and inhaled through the lungs by the intended victims. However, if the agent is properly selected and used the consequences could be devastating. A well-known calculation by the US Congress Office of Technology Assessment in 1993 suggested that whereas a 1Mt nuclear weapon might kill up to 1.9 million people if detonated over Washington, DC, 100 kg of anthrax spores in the right conditions could kill up to three million (Office of Technology Assessment, 1993). There are a number of other calculations in the open literature which essentially come to the same conclusion about the potential consequences of the use of biological weapons. These calculations have not been contested and have to be accepted as basically correct.

In order to achieve such effects, however, the agent selected would need to have certain characteristics. It would, for example, need to produce a certain effect consistently, be effective at a low dose, have a short and predictable incubation period, be easily produced on a large scale and be stable in storage and on dissemination. Clearly these are stringent requirements and far from all possible pathogens meet them. However, a good number of micro-organisms have been found to be more than adequate. Indeed the weapons designer is able to choose, for example, whether to use a lethal agent or one that merely incapacitates, or one that is not infectious after first use or one that remains contagious from the first (and further) victims. Obviously if an agent is selected which remains highly contagious there would have to be very effective means of immunisation available for friendly troops and populations in order for them not to be at risk from use of the agent. Such contagious agents were not weaponised in the US mid-century programme, but both plague and smallpox appear to have been weaponised on a large scale in the late Cold War programme of the former Soviet Union (Steinbrunner, 1998).

Agents

Because of the relative openness of US society a good deal is known, in general terms, about the agents that were weaponised for possible direct use against human beings in their offensive programme. Some details about these agents are set out below.

Anthrax

Anthrax is a disease of herbivorous animals caused by the bacterium *Bacillus anthracis*. Animals are infected from contaminated material in the field and the bacterium grows quickly in the body of the animal. When the animal dies the bacterium is able to form very environmentally resistant spores which can then be reactivated inside the body of the next victim. Humans can catch the disease through inhalation of the spores (pulmonary anthrax) or through lesions in the skin or by ingestion of contaminated material.

An infection through the lungs, if not treated rapidly by modern means, is almost inevitably fatal. The incubation period is some two to three days and death is likely in 95–100 per cent of untreated people within a few days afterwards. The effectiveness of available vaccines against a heavy inhalation dose in man is not known (Franz *et al.*, 1997).

These characteristics, particularly the environmental stability of the spores, make anthrax the choice agent in most known offensive biological warfare programmes. If the agent is dried and milled to the correct (approximately five microns) size to go into the lungs and remain there, and if an attack is carried out at night to avoid UV light degradation of the viability of the spores, anthrax would clearly be extremely effective (Pile *et al.*, 1998).

Brucellosis

Brucellosis is a bacterial disease in domestic animals. Four types of brucellosis can also attack human beings: *Brucella melitensis*, which attacks goats; *Brucella abortus*, which attacks cattle; *Brucella suis*, which attacks pigs; and *Brucella canis*, which attacks dogs. The severity of the disease in humans is worst with *Brucella melitensis*.

The disease is not often lethal in humans but results in long-term incapacitation, which can last for months and recur again later. There is no vaccine available for humans and the organism is capable of surviving for long periods in the environment.

Q-Fever

This is a widespread disease of domestic livestock caused by the obligate intracellular (rickettsial) organism *Coxiella burnetti*. Humans infected are not likely to die but suffer an incapacitating illness. The organism can persist for long periods in the environment and is remarkably infectious.

Tularemia

Tularaemia is a natural disease of rodents, rabbits and hares. The causative agent's *Francisella tularensis* can also infect human beings in a variety of ways, for example through contaminated food or through the lungs. The organism is hardy in the environment and infects at low dosage. The agent weaponised by the United States produced a debilitating illness but could also kill about a third of those infected if they were not given modern medical treatment.

VEE Virus

Venezuelan Equine Encephalitis (VEE) Virus is a member of the Alphavirus genius of the Togavindae. In nature it is normally carried by an arthropod biting vector. The virus has a wide range of natural hosts. It produces a fatal disease in horses and an incapacitating influenza-like illness in humans. This lasts for about a week. An effective vaccine is available, but the virus is very stable in the environment and highly infectious to humans. It could also be produced in large amounts by relatively unsophisticated methods.

So it can be seen that during the middle years of this century a range of bacterial agents were weaponised by the United States. At that time more was known about these simple single cellular organisms than about the viruses. Viruses are not cellular organisms, they consist of genetic material and a protein coat and must invade a living cell and take over its operations in order to reproduce. The Q-fever organism *Coxiella burnettii* whilst being cellular, similarly has to invade a cell in order to reproduce. The other bacterial agents just have to invade the body in order to grow and produce disease. However, many bacteria produce deadly toxins, and this provided another possible way in which to use biological agents. As toxins are non-living chemicals, if they are capable of weaponisation a faster effect could be expected because they would not have to reproduce in the body to cause a disease. Two such toxin agents were weaponised by the United States.

Botulinum toxins

The bacterium *Clostridium botulinum* produces seven neurotoxins. Different strains of the bacterium produce these proteins which have a molecular weight of about 150,000. The toxins are labelled A through to G. All of the toxins act in a similar way and induce their effects whether inhaled or ingested. Many deaths have resulted from contamination of food by these toxins and Iraq is also known to have weaponised this agent.

Botulinum toxins are 15,000 times more toxic then VX nerve agents (the most deadly chemical weapons agent). In fact botulinum toxins are the most toxic compounds known for human beings with an estimated lethal dose (for serotype A) of just 0.001 ug/kg of body weight. The toxins act presynaptically at cholinergic synapses and prevent the release of the transmitter. This leads to progressive and rapid muscular paralyses, and death by respiratory muscle paralyses if the victim is not treated effectively. An antitoxin is available but the victim requires testing for horse serium sensitivity before administration.

Staphylococcal Enterotoxin B (SEB)

The bacterium *Staphylococcus aureus* produces a number of different exotoxins. One of these is SEB. As the toxins exert their effects on the victim's gastrointestinal tract in food poisoning they are, somewhat confusingly, called enterotoxins. SEB frequently causes food poisoning through contaminated food. Because it is toxic at very low doses SEB could also be used as a biological weapons agent through inhalation. The toxin causes incapacitation at doses much, much lower than would be required to kill and is therefore classed as is an incapacitating rather than a lethal agent.

SEB is a superantigen. It causes its effects through a complex interaction with the victim's immune system which leads to an over production of cytokimes. Reactions set in within hours of inhalation of the toxin and may lead to an illness involving fever, headaches, etc. lasting several days.

Most patients given supportive care could be expected to fully recover in a few weeks.

Quantities of agent

An important question in relation to the prevention of biological warfare is what is a militarily significant quantity of an agent such as anthrax, and could production of such an amount be detected. If we consider an attack at night against unprotected victims from a stand-off clandestine source it is possible to obtain a rough estimate using standard calculations (Bartlett, 1996). If ID_{50} is the dose required to infect 50 per cent of an exposed population it turns out that a single attack would require 10^{12} ID_{50} or *one trillion doses*.

Now the simplest method for a proliferator would be to obtain this quantity of agent by growing a bacterium by fermentation. Assuming that a concentration of 10^8 bacterial cells/ml could be reached in the fermentors, and that the agent was anthrax, which has an estimated ID_{50} of about 10^4 cells, the proliferator would need to grow:

$$10^{12} \times 10^4/10^8 \times 1000 = 100,000 \text{ litres of bacterial suspension}$$

That would require, for example, ten runs of ten fermentors each with an individual capacity of 100 litres. Together with the ancillary equipment needed that would require a not insignificant amount of space.

Effective weaponisation of the agent would also require drying and milling and a certain amount of testing even for an agent such as anthrax (Larson and Kadlec, 1996). For widespread use there would also have to be training and equipping of the troops who would be tasked with using the biological weapon. Such activities would be increasingly subject to detection by other countries.

Futures

The biotechnology revolution had an impact on the concerns of states even during the early 1980s. By the time of the Second Five Year Review Conference of the Biological and Toxin Weapons Convention in 1986 it was realised that genetic engineering might allow the much more straightforward production of large quantities of bacterial toxins.

At the present time the major proliferation concern remains concentrated on the 'classical' agents, such as anthrax, as it is assumed that these would not require a large amount of testing. However, it is clear that the offensive biological warfare programme of the former Soviet Union had begun to investigate very dangerous viral agents (Breman and Henderson, 1998) to a much greater degree than was possible (given the state of knowledge at the time) of the US offensive programme. The Soviet programme had also begun to apply genetic engineering methods, for example to increase the range of antibiotic resistance of plague.

If we consider the offensive biological warfare programmes of the First World War as the first generation of scientifically based programmes, and those at the time of the Second World War as a second generation, we might consider the programme of the former Soviet Union as being a third generation. Each generation of programmes represents the application of increasingly sophisticated scientific knowledge. This then raises the question of what might be the characterisation of a fourth generation programme as the biotechnology revolution develops and spreads over the next few years?

According to the US Department of Defence (Cohen, 1997) we could possibly expect to see: benign micro-organisms genetically altered to produce a toxin or bio-regulator; micro-organisms resistant to antibiotics, standard vaccines and therapeutics; micro-organisms with enhanced aerosol and environmental stability; microbiologically altered micro-organisms able to defeat standard identification, detection and diagnostic methods; and combinations of these four types of development.

As the revolution in biotechnology and molecular medicine develops, it is clear that a wide variety of other ways in which scientific knowledge could be misused will become available to the weapons designer. Unless the process is stopped this will lead to the development of further more terrible generations of new offensive biological warfare programmes in the next century (Dando, 1999; Nathanson, Darvell and Dando, 1999)

Conclusion

There is an urgent necessity for the biomedical community to take much greater responsibility for what is done with the knowledge it generates. The first requirement is for the level of awareness of the misuse of biological and medical knowledge in offensive biological warfare programmes over the last century to become common knowledge in the community today. Then we must take care to monitor and discuss development that could be a cause of concern and make our views on what should be done widely known. Whilst there is a role for the individual scientists, our professional organisations should also put this issue much more centrally on to their agendas.

Such a change in the biomedical community's stance will not be sufficient to prevent new developments in biological warfare but it seems to me to be a necessity in humanity's overall strategy. At the back of our minds we have to remember that many previous scientific revolutions so far have been used to refine weapons of war and we will have to work very hard to prevent the current revolution in biotechnology suffering the same fate.

11

Anti-crop biological warfare: implications for the Biological and Toxin Weapons Convention, and the Convention on Biodiversity

Simon Whitby

Introduction

Biological warfare can be defined as the deliberate use of bacteriological (biological) organisms against man, animals and plants. Anti-crop biological warfare can be defined as the deliberate use of naturally occurring plant pathogens (disease producing organisms), such as fungi, bacteria and viruses, and vectors associated with the transmission of these pathogens (nematodes and insects). Possible attacks on humans are increasingly being discussed in the literature (Wheelis, 1997) and we know that attacks were made against animal stocks by both sides in the First World War. But why attack plants?

A possible answer to that question appeared on ProMED-mail on the 6 August 1996. This was in an article titled 'Feeding a Hungrier World' from *Phytopathology News* (Anderson, 1996). This article began by arguing that new farming techniques and crop strains had averted in the 1990s the chronic food shortages that affected many countries in the 1980s:

> RICE: nine-tenths is grown in the Asia/Pacific region, where demands is projected to jump 70% in the next 30 years; yields rose by 52% and this crop is increasingly important in Africa and Latin America.

> WHEAT: is eaten by more very low-income people than any other crop; yields per acre nearly doubled from 1974 to 1994 in all regions, except Latin America. (Anderson, 1996: 90)

Despite the achievements of the 1990s the outlook is ominous. As the report continues:

> officials ... are doubting that scientific innovation alone can raise production sufficiently to keep pace with hunger. The planet's population is projected to grow by about 85 million people a year for two or three decades. Ninety percent of that will occur in the Third World, approximately doubling demand for food there by 2025. (Anderson, 1996: 90)

In such circumstances attacks on basic food crops, like rice and wheat, as an 'economic' assault on an enemy could look an increasingly attractive option in the war-torn regions of the world over the next few decades. Moreover, a wealth of knowledge and practical experience has already been applied in biological warfare programmes on targeting crops. Scientific techniques which open up new possibilities regarding the genetic modification of biological organisms extends the technical capabilities required in causing deliberate disease in food and cash crops.

It is argued in this chapter that biological warfare against crops must be taken seriously today and that efforts to strengthen the prohibition against it should be re-doubled. The efficient and effective implementation of a strengthened prohibition regime for biological weapons will complement existing initiatives that intend to promote the peaceful use of micro-organisms and the strengthening of biosafety.

The chapter is organised into reasonably distinct but related sections. It will be shown in the first section that the notion of attacking an adversary's access to food and cash crops is not without precedent. This form of warfare has a long a detailed historical lineage, and there is one example in the public domain of a mature anti-crop biological warfare research and development programme resulting in the standardisation of agents, and the deployment of munitions. Recent evidence suggest that the proliferation of this form of warfare is a particular concern for arms control and international security

In the second section there is a discussion regarding the ongoing initiative to strengthen the 1972 Biological and Toxin Weapons Convention (BTWC) and the way in which the BTWC verification and compliance protocol is envisaged to outlaw the development, production and stockpiling of naturally occurring and genetically modified plant pathogens of types and in quantities that have no justification for prophylactic, protective or other peaceful purposes.

On a separate track, international efforts to implement the Convention on Biodiversity have centred around the three-fold aim of: conservation of biodiversity, sustainable use of the plant's biodiversity and the equitable sharing of the benefits of genetic resources. The third section discusses the Biosafety Protocol and the UNEP (United Nations Environment Programme) International Technical Guidelines on Safety on Biotechnology which together establish appropriate procedures, including, in particular, advance informed agreement in the field of the safe transfer, handling and use of any living modified organism resulting from biotechnology which may have adverse effects on the conservation and sustainable use of biological diversity.

The final section will discuss the existing 1952 multi-lateral plant protection agreement and the *ad hoc* and non-governmental disease reporting mechanisms which will aid in the effective implementation of the protocol on the prohibited military use of micro-organisms under the BTWC and in the effective implementation of the Biodiversity Protocol under the Convention on Biological Diversity which is envisaged to work in parallel to the UNEP Guidelines on Biosafety.

A brief history of war against crops

The threat to major food crops was summarised by van der Plank in his 1963 textbook on *Plant Diseases: Epidemics and Control*:

> We often call an epidemic explosive. In time of peace the adjective is neatly descriptive. In time of war it could be grimly real in the military sense. An enemy has few explosives to surpass a pathogen that increases at the rate of 40% per day, compounded at every moment, and continues to increase for several months. Spores are as light as poison gas ... Many types of spore disperse as easily as smoke. Many are tough and durable. They have only to be dispersed in the proper places at the proper times. Nature sees to the explosion. (van der Plank, 1963: 212)

Despite its age, the US programmes of the 1940s, 1950s and 1960s provide a useful starting point for developing an appreciation of the way in which warfare against crops was planned in past offensive biological warfare programmes. During the course of the US programmes, extensive consideration was given to the use of a range of agents and some were weaponised.

The official US history summarised anti-crop biological warfare activities as follows:

> research on [anti-crop] BW agents included strain selection, evaluation of nutritional requirements, development of optimal growth conditions and harvesting techniques and preparation in the form suitable for dissemination. Extensive field testing was done to assess the effectiveness of agents on crops. (Laughlin, 1977: C-2)

A recent study concluded that the US eventually standardised five anti-plant biological agents. The identity of two of these agents is not known, but the study concluded that amongst those considered may have been late potato blight (*Phytophthora infestans*) and bacterial soft rot (*Erwina caratovara*) which affects carrots, potato, cabbage and onions. Here, however, we will concentrate our attention on two of the three known agents: those intended to attack wheat and rice.

Not only are such targets of critical importance today, but so are the pathogens worked on. The US chose stem rust of wheat (*Puccinia graminis tritici*) as one of its principal anti-crop agents. Writing in the early 1960s, van der Plank suggested (presumably without knowledge of the US work):

> Stem rust of wheat is an obvious choice as a weapon. It has long been known as the disease which when it is rampant destroys more in less time than any other ... Great researches, unsurpassed in plant pathology, have uncovered just how stem rust spreads and when it spreads. A long series of publications ... tell

how wind, sunshine, temperature, rain, dew, and humidity of the air govern the dispersal of stem rust over a million square miles of North America ... (van der Plank, 1963: 213)

Plans to attack wheat crops

A study released under the US Freedom of Information Act in 1963 discusses the potential vulnerability of the food economy of the former Soviet Union and its European satellites (Perkins, McMullen and Vaughan, 1958). It begins with an analysis of the impact of various levels of reduction of calorific intake on body weight. For example, it is estimated that a 50 per cent cut in intake over a period of twelve months would result in a 30 per cent reduction in body weight given base-level intakes just after the Second World War. Then the economic, social and political impacts of various reductions in body weight are estimated. For example, a 10 per cent reduction in body weight is estimated to be the likely cause of serious civil disorder and strife.

Estimates are then made of the average calories per capita per day in the former Soviet Union. This intake is then broken down by food types and it is concluded that 72 per cent of the calorific intake per capita per day was provided by grain. The study therefore concludes that 'a large fraction of the diet is threatened if wheat can be successfully attacked' (Perkins, McMullen and Vaughan, 1958: 1). The study then goes on to discuss in some detail the characteristics of stem rust of wheat, *Puccinia graminis tritici*, which make it ideal for such an attack.

In spite of the absence of information relating to the means of dissemination in this study, we can, nevertheless, follow earlier thinking on the operational use of this agent from a 1950s document titled, 'Feathers as Carriers of Biological Warfare Agents: I. Cereal Rust Spores' (Biology Department, 1950: 1), which was declassified in 1977. This study begins by reviewing earlier tests in which it was shown that birds contaminated with spores of *Puccinia graminis avenae*, Race 8 could be used to infect seedling oats effectively. This, of course, in not an illogical potential route of infection.

A first test demonstrated that heavy infection resulted from releasing birds dusted with spores into cages in test fields for one to twenty-four hour periods. In a second test homing pigeons dusted with spores flew 100 miles before being put into cages in the test fields. Again primary infection of the crop resulted, indeed viable spores remained on the birds in large numbers nineteen days after their flight. A third test involved releasing contaminated birds from aircraft and then their remaining on a test field for two hours. Heavy infection of the crop again resulted from the test. The study concludes, however, that: 'The problems involved in collecting, stockpiling, and processing birds would tend to limit the use of this method for the wide-scale distribution of cereal rust spores over an enemy territory' (Biology

Department, 1950: 4). The fact that feathers were very efficient at holding large quantities of spores suggested the use of feathers alone. Thus: 'The purpose of the feathers test described in this report was to determine if feathers dusted with cereal rust spores and released from aircraft in a M16A1 cluster adapter (used for leaflets and fragmentation bombs) will permit the transference of spores to cereal plants so that rust infection may ensue' (Biology Department, 1950: 4). Test drops of the bomb showed that feathers released from an altitude of 4,000 feet could spread over an area of twelve and a half square miles.

The target area for the test was approximately eleven miles long and one and a half miles wide. In this area fifteen half-acre plots were planted with the Overland variety of oats which was susceptible to the test agent, but resistant to other common races of cereal rust. Three cluster bombs were released one mile upwind of the target area and the resulting infection carefully assessed over the following weeks. Eleven plots were initially infected and eventually all the rest were infected by secondary distribution of spores on the wind. It was estimated that the yield reduction in the field initially infected would be in the region of 30 per cent.

The report concludes that 'feathers dusted with 10 per cent by weight of cereal rust spores and released from a modified M16A1 cluster adapter at 1300–1800 feet above ground level will carry sufficient numbers of spores to initiate a cereal rust epidemic' (Biology Department, 1950: 5). However, it was also found that spore distribution was much wider (twenty-five square miles to 7.7 square miles) than feather distribution. Whilst the report does not state it, this implies that spores alone could be an effective means of distribution.

It was estimated that anti-crop biological warfare agent fills for dissemination by the above feather bomb could be produced by the spring of 1951, and the agent fill was standardised by the Chemical Corps Technical Committee on 25 May 1951. Arrangements were made for some 18,000 of the 70,000 munitions in storage to be modified by the Chemical Corps to make them suitable for use. In connection with this development, according to Miller (1952) a contract had been established with General Mills. The above munitions were intended for delivery by piloted aircraft.

An Air Material Command Plan (12–53) for the delivery of 3,200 of the above munitions had been devised by 1953. The period of operations for the delivery of this munition would be limited to the following: 1 March through to 30 May each calendar year. By October 1954 the causal agents of stem rust of wheat and rye were available in refrigerated storage in quantities sufficient to meet Air Force requirements for anti-crop biological warfare. In accordance with Presidential approval, the USAF Headquarters would issue a directive to the Chemical Corps to transfer agent filled munition components to aircraft for transportation to overseas bases. The agent filled munition components would subsequently be transferred to some 4,800 M115 anti-crop warfare biological bombs that had been pre-positioned at two over-seas bases. At no time was the munitions fill stored outside of the United States. According to Miller: 'Therefore, the Strategic Air Command had the capability of conducting biological anti-crop operations as part of a strategic air offensive if directed' (Miller, 1952: 110).

The E95 self-dispersing biological bomb
A further example of research and development into munitions is outlined by Miller (1952). Testing with the M115 feather bomb had revealed that dissemination of anti-crop biological warfare agents in 'pure form' had proved to be a more effective means of dissemination than dissemination of anti-crop pathogens via the medium of feathers. Initiation of the E95, 10 ounce, self-dispersing biological bomb that was intended for the dissemination of dry pathogens, followed development of the M115. Some of the characteristics of this munition are described by Miller, as follows:

> The E95 was to be clusterable in a hopper designed to fit into the bomb bay of bomber aircraft, and was also to be capable of being stowed in and released from external wing carriers of fighter aircraft. A major development problem proved to be the design and procurement of a simple and inexpensive fuze to insure munition operation prior to impact. (Miller, 1952: 99)

The E86 anti-crop biological warfare bomb
The US Army Chemical Corps and the US Air Force also considered the E86 bomb as a means of delivery for anti-crop biological warfare agents. The E86 was based on the same operational principles as the M115 bomb but was larger at 750lb. Designed for the dissemination of dry anti-crop agents, the E86 munition was intended for delivery from the external carriage on piloted aircraft such as the B-47 and B-52. Development of this munition began in October 1951, when in October 1952 production was assigned under contract to the Ralph M. Parsons Company. Procurement was initiated for some 6,000 E86 cluster munitions in Fiscal Year 1953, and production of munitions for operational use was not expected prior to January 1958.

The E77 anti-crop warfare munition
Work began on the E77 anti-crop biological warfare munition in 1950. According to Miller: 'Preliminary military characteristics were accepted by the Chemical Corps Technical Committee in April 1951, and were revised in November 1952' (Miller, 1952: 112). This munition was distinct from the E73 and the E86 anti-crop biological cluster munitions in that it was designed for the dissemination of dry anti-crop biological warfare agents by free-floating un-manned balloons and not for dissemination by piloted aircraft. According to one source this munition copied the method developed by the Japanese during the Second World War which resulted in numerous attacks on the US mainland by Japanese balloon bombs. Positioned beneath

the hydrogen-filled balloon was a gondola containing a consignment of five biological containers. The following associated mechanisms were also carried in the gondola: a heating mechanism; a mechanism designed to eject the munitions payload – consisting of feathers and anti-crop agent – from the five biological containers; and a mechanism designed to neutralise the payload to prevent its dissemination over friendly territory.

According to Miller the E77 biological munitions system was designed as a strategic weapons system:

> It was to be launched by a special group assigned or attached to the theatre air commander. Five launching sites were planned. Training was to be the sole responsibility of the 1110 Air Support Group, Headquarters Command, stationed at Lowry Air Force Base, Colorado. A hard core of personnel could be expanded if necessary. Air and Airways Communication Center and Air Weather Service were to cooperate in tracking [the munitions]. (Miller, 1952: 113)

Initial assessments with regard to the quantity of agents required to achieve a specific area coverage proved unrealistic and the Air Force found that in order to successfully deliver 150 functioning munitions, some 2,400 munitions would have to be launched during a single season requiring distribution of some 4,000 munitions amongst five launching sites.

Plans to attack rice crops

A British document of late 1945 notes that: 'Several fungi known to cause serious disease and destruction of crop plants have been studied in America as possible means of attack of enemy crops.' The fungi chiefly considered were few in number, including: 'Agent IE *Piricularia oryzae* Br. and Cav. Some varieties of rice are very susceptible and others resistant to attack by this fungus. The resulting disease has been reported to be serious in Japan, China and India, but it has not been of much importance in the USA' (Joint Technical Warfare Committee: 1945).

We can gain an insight into this early work from a paper published in *Phyotopathology* in 1947 which reported studies carried out at Camp Detrick under laboratory conditions and field testing in Texas (Andersen, Henry and Tullis, 1947: 281). The paper covers issues such as varietal resistance, the effect of the age of the plant on the susceptibility to infection, environmental factors affecting primary and secondary infection, the longevity of conidia and the susceptibility of other cereals. It was found for example that seedling plants were susceptible to attack by the fungus, then susceptibility declined with increasing age until the plant again became more susceptible as the seed heads began to form.

It is interesting to note that E. C. Tullis, one of the authors of the 1947 paper, was also the senior author of a major 1958 study (Tullis *et al.*, 1958) published by Fort Detrick under the title, 'The Importance of Rice and the

Possible Impact of AntiRice Warfare'. This case study, which was declassified in its entirety in April 1961, provides a primary insight into thinking about how best to attack plant crops, and therefore also into the particular characteristics that such an attack might have. The 185 pages are divided into a number of sections and sub-sections which are of particular interest (see table11.1). It is obvious even from an initial perusal of these headings that anti-plant economic warfare is unlikely to be a small-scale *ad hoc* enterprise undertaken without careful planning and preparation. Any signs of such planning and preparation would, of course, be an initial indicator of a difference between a deliberate and a natural outbreak.

Table 11.1 Rice and the possible impact of anti-rice warfare

Section I	*Introduction*	
Section II	Rice	(A) Historical background. (B) Plant characteristics. 1. Varieties. 2. Climate. 3. Germination. 4. Water requirements. 5. Height of Plant. 6. Tillering. 7. Maturation period. (C) Culture of rice. 1. Various countries and methods. 2. Yields. 3. World acreage and production. (D) World trade. (E) Dietary aspects. 1. Availability, Consumption and preference. 2. Nutritional value of rice. (F) Milling, Products and utilisation.
Section III	Possible anti-rice agents	(A) Chemical characteristics. 1. Agent effectiveness. 2. Industrial production. 3. Area, coverage and capability. (B) Biological characteristics. 1. Rice blast.
Section IV	Possible targets in China	(A) General rice-producing areas. 1. Central rice, wheat. 2. Szechwan Basin. 3. South-eastern rice-tea. 4. Southern double-cropping. 5. South-western area.
Section V	Conclusion and recommendations	

The 1958 study states that: 'Famine ... provides a real threat; its terrible consequences have been experienced in most Eastern countries, especially China. An ability to induce famine by antifood warfare can provide a strong deterrent to aggression ...' (Tullis *et al.*, 1958: 2). Moreover, the fact that this study was produced when the US anti-plant programme was under threat does not negate the insight it provides to the planning and preparations being undertaken. The overall 'logic' set out in the study is summarised in table 11.2. Whilst it deals with both chemical and biological attacks on rice the study argues that these are not at variance: 'These types of agents are considered complementary and not competitive. The chemical agent can be used in climatological or environmental situations unsuitable for the pathogen, however, where the situation is favourable for the disease its use would offer a means of attacking huge concentrations of rice ...' (Tullis *et al.*, 1958: v). Nevertheless, the amounts of agent needed are very different, '225 grams per acre for the chemical versus 0.1 to 0.2 grams per acre for the pathogen'.

Table 11.2 The logic of anti-rice warfare against China

1	Only 12 per cent of the land is suitable for agricultural production, and only 3 per cent is used to grow rice.
2	Rice production is concentrated in two regions – Shanghai and Canton – and thus is particularly susceptible to attack.
3	Rice contributes approximately 60 per cent of the calorific content of the diet.
4	An anti-rice warfare campaign may achieve between 33 and 80 per cent reductions in calorific intake.

Section II of the study gives a very detailed account of what was known of rice and how it was grown in China at that time. In regard to varieties and their resistance to attack the study states that:

some hundred of rice variants have been examined and the results so far have *not* indicated that a different race of an organism would be required in order to produce a disease on each of the rice variants ... it appears at present that a broad spectrum of rice variants will be found susceptible to a few strains of the organism. [Emphasis added] (Tullis *et al.*, 1958: 10)

Three isolates of *Piricularia oryzae* from Costa Rica (825), the Philippines (920) and Nicaragua (640) had been tested for resistance and susceptibility out of 483 rice varieties from different parts of the world. The study quotes a report of the Eleventh Tripartite Conference which stated, 'Currently 14 distinct races have been identified on the basis of susceptibility of a number of varieties of rice plants' (Tullis *et al.*, 1958: 1). The report went on, 'by using mixtures of races it is hoped that virtually all oriental varieties can be successfully attacked by this agent' (Tullis *et al.*, 1958: 1). Therefore, it would appear that a serious attacker would have to use a mixture of different types of the pathogen in an effort to ensure the decimation of the food crop. Such a complex assault could again suggest that the outbreak was not natural.

So far then it appears that an attacker with serious intent to do severe damage to a staple crop and in order to inflict meaningful economic damage on an enemy would have to prepare and plan the assault with care. It also seems probable that to ensure a successful result a complex mixture of different types of pathogen might be used. Another feature that might be expected from this line of argument is the scale of the assault. Whilst the study only gives details for the use of the chemical agent the areas needing to be attacked to achieve significant results are very large. This contention is born out by the estimated number of aircraft sorties (with Navy Aircraft and Aero 14B Spray Tanks) required for the delivery of anti-crop chemical agents on to target crops in the region in the vicinity of Shanghai (approximately 400 sorties required to attack an area of some 2,086 square miles), and by the estimated number of aircraft sorties (again with the above means of delivery) in the region in the vicinity of Canton (estimated to be approximately 1,997 required to attack an area of some 10,600 square miles). Although the amount of agent required on target favoured the pathogen (at only 0.1 to 0.2 grams per acre) when compared with the chemical anti-crop agent, the area coverage for agents delivered to the above targets would suggest that similar large-scale operations were envisaged for the delivery of anti-crop biological warfare agents.

In conducting anti-crop warfare the US Navy attached particular significance to the advantages enjoyed by carrier-based aircraft, arguing as follows:

> In many areas the growing of rice virtually borders the seacoast thus presenting short distances to targets where the adjacent waters are subject to Naval supremacy. A high proportion of flying time could be devoted to spraying the crops. Minimised fuel requirements per sortie and turn-around time would contribute to a most efficient operation. (Tullis *et al.*, 1958: 114)

It would seem reasonably clear that the above preparations in the 1950s would have allowed the US to conduct large-scale operations in anti-crop biological warfare. Given the weapons systems involved at that time, an attack would surely have been detectable. However, given the nature of the pathogen and the subsequent development of 'weaponless' dispersal (for example. through a spray system, such as the Navy Aero A1, and A2) evidence that an attack on a food crop had taken place may not have been obvious until the disease had become well established on the crops.

There is no evidence which suggests that the US ever deployed its anti-crop biological warfare capability, and further research and development into agents and munitions was halted in the late 1960s when Nixon unilaterally renounced research and development into offensive biological warfare. Yet, more recently, the science and technology of biological warfare against crops has been given a new lease of life with Congress approving $23 million for an anti-narcotics programme which includes research into plant-specific pathogens. Amongst the proponents of the use of pathogens against plants that produce narcotics, such as cocaine, heroine and marijuana, the identification of such strains has recently been hailed as a so-called 'silver bullet' in the war against drugs. In this connection, opposition to the use of such pathogens has been raised on a number of levels. First there is the concern that plant-specific pathogens might, in some circumstances, spread to other plants. Second, the science and technology of anti-narcotics programmes, conduced as part of bi-lateral or multi-lateral co-operative aid programmes, could be used to target drug-producing regions without the consent of the states concerned. Finally, and of greatest concern, is that the science and technology of a capability to target crops with plant-specific pathogens – which may or may not have been subject genetic manipulation – will inevitably result in a wealth of knowledge that could be readily applied to the pursuit of a capability to wage war against food and cash crops.

In addition to the above developments in anti-crop biological warfare, concern has been raised over the proliferation of this form of warfare. Recent reports have claimed that Russia had conducted extensive work on acquiring a capability to wage war against crops (Rufford, 1998), and a report released in 1995 by the United Nations Special Commission (UNSCOM) on Iraq revealed that Iraq had worked on anti-crop biological warfare which involved consideration of the fungus of the genus *Tillitia* which causes cover smut or bunt of wheat probably as a means of attacking staple crops in Iran (Whitby and Rogers, 1997: 305).

Given the extent of the US anti-crop biological warfare programme and current concerns over the proliferation of this form of warfare it would therefore seem appropriate to consider the current status of the international prohibition which bans deliberate disease against plants.

The BTWC and the use of plant pathogens for peaceful purposes

The Biological and Toxin Weapons Convention (BTWC) came into force in 1975, with Russia, the UK, and the USA as co-depository states. The BTWC bans the production, stockpiling or retention of:

> [m]icrobial or other biological agents, or toxins whatever their origin or method of production, of types and in quantities that have no justification for prophylactic, protective or other peaceful purposes.

The absence of a formal mechanism to monitor the compliance of States Parties has reduced the 1972 Biological and Toxin Weapons Convention to a gentleman's agreement undermining its ability to resolve allegations of non-compliance and to address growing concerns about the proliferation of biological warfare capabilities (Tucker, 1998: 20).

Since 1994, an Ad Hoc Group, meeting at the United Nations in Geneva, has been mandated to *inter alia*:

> consider appropriate measures, including possible verification measures, and draft proposals to strengthen the Convention, to be included, as appropriate, in a legally binding instrument [a compliance and verification Protocol], to be submitted for the consideration of the States Parties

Requirements for declarations, procedures for visits to facilities, provisions for investigations together with safeguards for confidential information have emerged from the negotiations as key and essential elements of a Protocol. As the negotiations approach the end-game there exists the real possibility that agreement on the Protocol will be reached over the next 12 to 18 month period (correct at time of writing) although the final details have yet to be negotiated.

During the course of the Protocol negotiations plant pathogens of importance to the BTWC have been listed in relation to efforts to attempt to identify a mechanism in the Protocol by which verification of the BTWC could be achieved. It must be noted that the listing of such pathogens was not intended to prejudice the positions of delegations on the issues under consideration in the Ad Hoc Group, and, according to one commentator, confining a list of pathogens to the Protocol, 'makes it quite clear that these are aids to the present implementation of the Convention and in no way limit the General Purpose Criterion [which allows the BTWC to prohibit the application of any biological and toxin agents as] embedded in Article I of the Convention' (Dando, 1999: 14).

Nevertheless the plant pathogens identified during the course of the negotiations include those that are known to have been selected as anti-crop agents in past biological warfare programmes. Moreover, a number of plant pathogens have been identified as possible candidates for genetic manipulation.

At the Sixth Session of the Ad Hoc Group, 3–21 March 1997, a Working Paper (BWC, 1997: 2) produced by South Africa discussed a number of plant pathogens in relation to the following criteria:

1 Agents known to have been developed, produced or used as weapons.
2 Agents which have severe socio-economic and/or significant adverse human health impacts, due to their effect on staple crops, to be evaluated against a combination of the following criteria:
   • Ease of dissemination (wind, insects, water, etc.);
   • Short incubation period and/or difficult to diagnose/identify at an early stage;
   • Ease of production;
   • Stability in the environment;
   • Lack of availability of cost-effective protection/treatment;
   • Low infective dose;
   • High infectivity;
   • Short life cycle

Out of a possible 20 plant pathogenic organisms the Working Paper identified a number of plant pathogens, and the diseases with which they are associated, as potential anti-plant biological warfare agents (see table 11.3). Important potential BW agents against staple crops, such as wheat (stem or black rust, cover smut or common bunt), rice (bacterial blight) and a wide range of vegetables and beans (cottony soft-rot or white mold etc.), are those that are considered to be candidates for genetic manipulation in order to increase their potential. It has also to be said that the target organisms – the plants producing our staple foods – are becoming more and more intensively studied at the genome level. This brings both hopes of greater food production to meet human needs – the starting point of our discussion – but also a much more sophisticated understanding of their vulnerabilities to pathogens.

Whilst the BTWC prohibits the initiation of deliberate disease it is also intended to foster greater scientific and technological co-operation between Parties to the Convention. Such co-operation is intended to improve transparency and enhance confidence that pathogenic organisms are being used safely and for permitted purposes (Pearson, 1998b: 32). Article X of the BTWC promotes the peaceful use of micro-organisms and toxins, whatever their origin, and declares that States Parties to the Convention will:

> undertake to facilitate, and have the right to participate in, the fullest possible exchange of equipment, materials, and scientific and technological information for the use of bacteriological (biological) agents and toxins for peaceful purposes. Parties to the Convention, in a position to do so shall also co-operate in contributing individually or together with other States or international organisations to the further development and application of scientific discoveries in the field of bacteriology (biology) for the prevention of disease, or for other peaceful purposes.

In the light of two decades of scientific and technological developments in the fields of biology and biotechnology the Final Declaration of the Fourth Review Conference in 1996 stated further in this connection that:

> The Conference once more emphasised the increasing importance of the provisions of Article X, especially in the light of recent scientific and technological developments in the field of biotechnology, bacteriological (biological) agents and toxins for peaceful applications, which have vastly increased the potential for co-operation between States to help promote economic and social development, and scientific and technological progress, particularly in the developing countries, in conformity with their interests, needs and priorities.

A tension exists between the promotion of science and technology for peaceful purposes and the prohibition of agents and munitions that have '*no justification for protective, prophylactic or peaceful purposes*' (Rogers, Whitby and Dando, 1999: 67). The Protocol, it is argued, 'should not be a tool by which un-regulated advanced technology that could be used to create weapons of war is transferred from have to have-not states, but neither should it be used to prevent the legitimate transfer of technology for peaceful purposes' (Rogers, Whitby and Dando, 1999: 67).

Where developing countries have tended to view the implementation of Article X in economic rather than security terms, that is, as a framework for the transfer of scientific and technological know-how from north to south, industrialised countries have tended to see the benefits of Article X in terms of co-operation on biosafety and public health, particularly with regard to improved epidemiological surveillance. However, concerns over the peaceful use of micro-organisms are being addressed elsewhere and, in order to avoid unnecessary duplication with international activities that are of relevance to the strengthening of the BTWC, such initiatives must be identified and considered in more detail.

The convention on biodiversity and the biosafety protocol

International initiatives to develop controls on the products of biotechnology have continued in parallel to the BTWC and have centred around efforts to strengthen biosafety through the Convention on Biological Diversity (CBD), and the United Nations Environmental Programme (UNEP) International Technical Guidelines on Safety on Biotechnology.

The harmonisation of national and regional regulations in respect of the safe handling, transfer and use of micro-organisms with such international initiatives will improve transparency and build confidence that such organisms are being used safely and for permitted purposes. In respect of the development of national initiatives to promote the safe and permitted use of micro-organisms, according to one commentator: 'It is apparent that in many countries, information is already being collected and submitted to national authorities about activities and facilities handling, using or transferring dangerous pathogens with inspection being carried out to confirm that the regulations are being complied with' (Pearson, 1998b: 32). However, although national implementation legislation has been enacted in the European Community, the UK, the US and other countries, a detailed discussion of national and regional regulations is outside of the scope of this chapter which focuses only on developments at the level of international activity.

The Convention on Biological Diversity entered into force in December 1973. By February 1999 there were 175 Parties to the CBD, but seven including the US had not been ratified (Tansey, 1999). Article I of the CBD sets out the Objectives of the Convention as:

> the conservation of biological diversity, the sustainable use of its components and the fair and equitable sharing of the benefits arising out of the utilization of genetic resources, including by appropriate access to genetic resources and by appropriate transfer of relevant technologies, taking into account all rights over those resources and to technologies, and by appropriate funding.

Scientific and technological advances in biotechnology and genetic engineering are therefore of relevance to the CBD.

Article II of the CBD defines 'biotechnology' as follows:

> Biotechnology means any technological application that uses biological systems, living organisms, or derivatives thereof, to make or modify products or processes for specific use.

Subsequent review conferences have monitored the implementation of the Convention. Following a decision made by the first Conference of the Parties to the CBD in 1994 which mandated an Open-ended Ad Hoc Group of Experts on Biosafety to consider the 'need for and the modalities of a protocol', and, 'existing knowledge, experience and legislation in the field of biosafety', the second Conference of the Parties of the CBD in Jakarta 1995 decided that safety in biotechnology would be addressed:

> through a negotiation process to develop in the field of safe transfer, handling and use of living modified organisms, a protocol on biosafety, specifically focussing on trans-boundary movements, of any living modified organism resulting from modern biotechnology that may have adverse effect on the conservation and sustainable use of biological diversity, setting out for consideration, in particular, appropriate procedures for advanced informed agreement. (United Nations, 1995a: 1)

The Biosafety Protocol consists of some forty-three Articles together with Annexes. Whilst agreement remains to be reached (correct at time of writing) before the Protocol can be effectively implemented its provisions in respect of the safe transfer, handling and use of Living Modified Organisms (LMOs) under the Objectives of the Protocol as stated in Article I, and arrangements relating to Advanced Informed Agreement in respect of the trans-boundary movement of LMOs as stated in Articles 3 to 10 represent important measures which, once implemented, will help build confidence nationally, regionally, and internationally that novel organisms are being handled, transferred and used in safe ways and for permitted purposes. Such mechanisms are also of relevance to the efficient and effective implementation of a BTWC compliance Protocol.

On a separate track, United Nations Environmental Programme (UNEP) Guidelines on Safety on Biotechnology were accepted by the Conference of the Parties to the CBD as an interim mechanism for strengthening biosafety during the development of a Biosafety Protocol. The UNEP Guidelines resulted from a joint initiative taken by the UK and the Netherlands in the wake of the 1992 Rio Summit. The Introduction to the Guidelines which comprise some six chapters and seven Annexes states that:

> The Guidelines address the human health and environmental safety of all types of applications of biotechnology, from research and development to commercialization of biotechnological products containing or consisting of organisms with novel traits. (United Nations, 1995b: 1)

The UNEP Guidelines include the following provisions: Chapter II addresses General Principles and Considerations; Chapter III addresses issues related to Assessment and Management of Risks; Chapter IV sets out provisions for the establishment of Safety Mechanisms at National and Regional Levels which emphasise the importance of setting up and/or strengthening *'national and/or regional authorities/national institutional mechanisms for over-*

sight and/or control of the use of organisms with novel traits'. Chapter V sets out provisions for Safety Mechanisms at the International Level using Information Supply and Exchange which states that countries, *'are encouraged to participate in the exchange of general information about national biosafety mechanisms ... [and that] this form of information exchange can be carried out through direct information exchange, as well as through the creation of an international register or database'*. Finally, Chapter VI of the UNEP guidelines addresses issues related to Capacity Building. The UNEP guidelines are seen as complementary to the provisions for safety in biotechnology under the Biosafety Protocol and also contribute to the building of confidence nationally, regionally and internationally that such organisms are being handled, transferred and used in safe ways for permitted purposes.

Further international provisions for the safe and peaceful use of microorganisms are of relevance to strengthening the BTWC. Such provisions are set out in the International Plant Protection Convention. Additionally the role of non-governmental and *ad hoc* plant disease surveillance and monitoring activities will also be addressed.

The IPPC, and further surveillance and disease reporting mechanisms

The International Plant Protection Convention (IPPC) is a multi-lateral treaty that came into force in 1952. The Convention was deposited with the United Nations Food and Agriculture Organisation (FAO) and adopted by the FAO at a conference in the early 1950s. The objective of the IPPC is to: 'secure common and effective action to prevent the spread and introduction of pests of plants and plant products and to promote measures for their control' (Wheelis, 1999: 9). A 1997 amendment to the IPPC defines what is meant by the term 'pest' as follows, as: 'any species, strain or bio-type of plant, animal or pathogenic agent injurious to plants or plant products' (Wheelis, 1999: 9).

Under a 1997 amendment to the Convention, States Parties are obliged to establish national plant protection organisations with responsibilities that promote the development of surveillance and international information-sharing activities, including a mechanism where information relating to the occurrence, outbreak or spread of plant pests is deposited with the United Nations Food and Agriculture Organisation's (FAO) Commission on Phytosanitary Measures.

Under Article VIII of the IPPC, regional co-operation is promoted through the establishment of the following eight Regional Plant Protection Organisations: Asia and the Pacific; Australia and Pacific Islands; the Caribbean; Central America; European and Mediterranean; North America; Northern South America; and Southern South America.

According to Wheelis 'the requirements under the IPPC to share information on important plant diseases with other Member States should insure a high degree of transparency with respect to plant diseases of concern to the BTWC' (Wheelis, 1999: 13). Additionally, non-governmental and *ad hoc* activities regarding the monitoring and surveillance of the outbreak of disease in crops are also of relevance to the strengthening of the BTWC. Amongst the most prominent of these international surveillance and disease-reporting mechanisms for plants is the Program for Monitoring Emerging Diseases (ProMED-mail), an e-mail based service which provides extensive world-wide coverage of plant disease outbreaks.

Conclusion

The efficient and effective implementation of a strengthened prohibition against biological weapons and the current initiatives to strengthen biosafety, including mechanisms for international co-operation and improved surveillance, will contribute to the building of international confidence that organisms which pose a risk to plants and biosafety are being used safely and for permitted purposes.

Providing the prohibitions and controls on dangerous pathogens are effectively implemented in countries participating in the above international regimes, it will become increasingly unlikely that anti-crop warfare would go undetected. This is particularly the case in view of the scale of the anti-crop warfare operations as envisaged during the course of the US anti-crop biological warfare programme. Even if an attack did not involve traceable weapon systems, there does seem to be a clear tension between the characteristics that a serious offensive biological warfare attack against crops would need to have and the chances of being able to carry out such an attack without a considerable chance of it being detected as a deliberate use of a pathogen.

In the case of clandestine attacks against crops such as those that might be mounted by terrorists, given the widespread practice of restricting crop production to mono-cultures of major food crops in North America and Europe, there is the danger that considerable damage to crops – and significant economic loss – could occur prior to the implementation of measures to prevent the further spread of the disease. However, the sophisticated agricultural extension services of the kind required for the early detection of disease outbreaks in crops involve resources that developing countries – increasingly dependent upon the production of single staple food crops – often lack.

Given concerns over the proliferation of anti-crop warfare capabilities, should the current initiative to strengthen the BTWC with a compliance and verification Protocol fail the world faces the prospect of a weak prohibition against a major group of weapons of mass destruction during a period of fundamental scientific and technological change. The consequences may included new and more devastating weapons some of which may pose a significant risk to the crops upon which the worlds citizens are increasingly reliant.

12

The future of biological weapons control

Jez Littlewood

Introduction

In this section on biological warfare Dando has detailed the current capabilities of biological weapons and assessed the likely impact of the biotechnology revolution on such weapons. Whitby has detailed one particular aspect of this problem by looking at plant pathogens, their historical development and dangers posed by them. In the following chapter Spear outlines a range of strategies to prevent use of biological weapons: denial, deterrence, detection, defence and destruction.

This chapter deals with one specific response of the international community to the problem posed by biological weapons and warfare, one that predates the biotechnology revolution: the prohibition on biological weapons and their use. The central element of this prohibition is the 1972 Biological and Toxin Weapons Convention (BTWC).[1] It has, however, one fundamental weakness, in that the obligations within the BTWC cannot, currently, be verified due to the formulation of the Convention and the obligations contained therein. This shortcoming has been pushed to the fore by the recognition that the Soviet Union (USSR) was in breach of the Convention despite being a depository government. Additionally, the case of Iraq and its biological weapons programme has flagged the problems posed by these particularly onerous weapons. The BTWC transcends the strategies outlined by Spear in the following chapter and as such these strategies can be seen as a response to the failings of the BTWC, in that states have devised alternative strategies, both singly and with other like-minded states, to deal with the possible development and use of biological weapons. Although these strategies will be maintained in the future, the best opportunity for success in the areas of denial and detection rest on the outcome of current attempts to strengthen the BTWC and address the shortcomings of the Convention. That is to say, to build these strategies on the existing international law that prohibits the development, production and stockpiling of biological weapons among States Parties[2] to the Convention.

The central elements of this chapter therefore deal with the BTWC. The chapter is divided into three sections covering: the history of biological weapons control, the problem of verification, and the emerging structure and potential outcomes. I argue that the verification and compliance protocol currently being negotiated in Geneva should consist of a range of integrated strategies if it is to address successfully the dual-use nature of knowledge and technology applicable to both biotechnology and biological weapons. Failure to address this fundamental issue will likely result in separate strategies to deal with the problem of biological weapons being implemented by different nations at different times, further compounding the lack of confidence in the BTWC.

The history of biological weapons control

Biological weapons and warfare (BW) evoke extreme revulsion. The international community has condemned the deliberate use of disease against human beings, animals and plants (United Nations, 1972). This condemnation and the prohibition against use thereof is not because of the potential impact of biotechnology, which will make BW more feasible, but has a long history (Dando, 1999; chapter 10, this volume). In the twentieth century the Geneva Protocol of 1925 prohibited the use of chemical and bacteriological (biological) weapons in warfare among the parties to the Protocol. This formed the basic prohibition on use of biological weapons in war. However, the number and extent of reservations to the Geneva Protocol by the contracting parties effectively make it a no first-use agreement. Many states, including the UK, retained the right to retaliate in-kind if attacked with biological weapons. As Europe and the world slid back to war in the late 1930s the fear of attack with biological weapons was very real. Thus, states began to research and develop biological weapons as a deterrent against attack with such weapons. In actual fact only one state was confirmed to have used biological warfare, Japan (Harris, 1994), although one recent allegation also points to the USSR having used biological warfare during the battle for Stalingrad (Alibeck and Handelman, 1999: 31).

Research and development in biological weapons continued after 1945 in the Cold War period, mainly by the superpowers. By the late 1960s biological and chemical weapons began to emerge from the shadow of nuclear weapons, particularly following the conclusion of the nuclear Non-Proliferation Treaty in 1968. Two publications at this time (United Nations, 1969; World Health Organization, 1970) concluded that biological and chemical weapons presented a threat to humanity. This coincided with the unilateral renunciation of all offensive biological weapons activities by the United States (US) and a proposal by the UK to negotiate a new convention prohibiting the development, production and stockpiling of biological weapons (Goldblat, 1971: 256). This new convention would supplement, and strengthen, the existing Geneva Protocol by effectively prohibiting the main-tenance of retaliation in-kind capabilities, essentially outlawing all aspects of biological weapons and warfare.

How did such a situation arise? The widespread perception at this time and the one promulgated in the public domain, was that biological weapons had little military utility. Thus, together with the moral revulsion against these weapons, military forces could forgo BW capabilities and, conse-quently, a multilateral agreement to prohibit the weapons could be agreed in a short period of time. This 'belief' in no military utility does not stand up to serious scrutiny. A Nobel laureate and biological warfare specialist in the US testified before Congress in 1969 that the continuation of an offensive BW programme would lead to the proliferation of such weapons and was, 'akin to our arranging to make hydrogen bombs available at the supermar-ket' (US Congress, 1970: 89). More recently, one of the US representatives on the WHO report, and adviser on BW issues to the Nixon administration, noted that in 1969 the US realised:

> that our biological weapons program was pioneering a technology that, although by no means easy to create, could be duplicated with relative ease, making it possible for a large number of states to acquire the ability to threaten or carry out destruction on a level that could otherwise be matched by only a few major powers. Our biological weapons program therefore risked creating additional threats to ourselves with no compensating benefit and was under-mining prospects for combating the proliferation of biological weapons. (Mesel-son, 1999: 4–5)

The unilateral renunciation by the US certainly facilitated the attempts to control BW, but it is clear from the above statements that biological weapons should not be considered as inadequate for military purposes. This view is reinforced by a statement by the UK prior to the negotiations that delaying agreement on a prohibition on BW was a danger in itself, insinuating that future developments pointed to greater potential for biological weapons (Goldblat, 1971: 291). By 1972 there was sufficient agreement among the international community to complete the negotiations on the BTWC and it became the first disarmament treaty for a weapon of mass destruction. At the heart of the Convention was a legally binding prohibition not to develop, produce or stockpile biological weapons or means of delivery for biological and toxin weapons (Article I). It required States Parties to destroy or divert to peaceful purposes all biological weapons and means of delivery prohibited by Article I (Article II). Additionally, it also prohibited States Parties from assisting any other state or organisation from acquiring a biological weapons capability (Article III).

In the current period the BTWC is criticised for failing to produce biolog-ical disarmament. Commonly quoted figures suggest that in 1972 four states had, or could quickly acquire, an offensive BW capability. By the mid 1990s estimates suggested ten or more states were suspected as having BW capabilities (US Congress, 1993a: 82). However, the most important element of the BTWC should not be overlooked despite the shortcomings in the Convention. As Falk noted: 'The advantageous situation of the BW legal regime should be appreciated – cultural revulsion tied to a framework endorsed by all major governments, political ideologies, and social institutions, that prohibits possession and development of biological weaponry, as well as their threat or use' (Falk, 1990: 263).

The above still holds even in the turmoil of the post-Cold War world. However, setting high standards of conduct for international behaviour without the means to ensure such standards are being adhered to by all States Parties is of limited practical effectiveness (Zilinskas, 1998: 195). As has been proved with the examples of Iraq and the Soviet Union, reality since entry into force of the BTWC has taught us that trust and confidence in the regime can only be achieved if implementation is verified (Mossenlechner, 1999: 1). The main shortcoming of the BTWC has always been its lack of an effective compliance and verification mechanism within the Convention and it is this issue that the States Parties are currently seeking to address with the negotiations for a verification and compliance Protocol.

The problem of verification

Arms control agreements are always a product of the international political culture and context of their time (Croft, 1996: 31), and the BTWC is no exception to this general rule. When the UK initially proposed that a separate and new Convention on biological weapons be agreed it considered verification to be impractical since it would entail inspecting every facility and laboratory in the world (Goldblat, 1971: 256). Other leading specialists, Lederberg among them, likened verification to requiring a 'single world state' (US Congress, 1970: 114). Thus, no formal mechanisms for verification were established in the Convention. It is important to recognise that this resulted in a Convention with a serious loophole, but the BTWC was agreed during a period when intrusive on-site inspection mechanisms to verify obligations were not acceptable to many states and, had such verification mechanisms been required, the Convention would likely never have been agreed.

At the core of the BW verification problem is the nature of the knowledge and technology required to develop and produce a biological weapon. In common parlance it is known as 'dual-use' technology, in that the knowledge and technology required has both civilian and military applications. This factor does not make the biological sciences nor biotechnology unique since the same dual-use issue applies to the chemical industry and chemical weapons and to a lesser degree the nuclear industry and nuclear weapons. What does make biotechnology and BW different from the other examples is that 'biotechnology is information intensive rather than capital intensive, and much of the relevant data are available in the published scientific liter-ature. For these reasons it is virtually impossible for industrialised states to prevent the diffusion of weapon-relevant information to states of proliferation concern' (US Congress, 1993a: 85). This difficulty was as relevant in the late 1960s as it is in the contemporary period.

Developments post-1975, following entry into force, led to a gradual re-evaluation of the Convention. By the 1980s, when confidence in the Con-vention was at a low ebb, some considered that the treaty was 'poorly done' (Finder, 1986: 13), primarily due to the lack of verification provisions, while others considered it was the best available outcome at that time (Westing, 1985: 632). The shortcomings were noted by the States Parties at both the First and Second Review Conferences, in 1981 and 1986 respectively. Mea-sures designed to improve confidence in the Convention were taken at the Second Review Conference in the form of politically binding (that is volun-tary) Confidence Building Measures (CBMs) (United Nations, 1986; Hunger, 1996). Progress in verification technologies and political breakthroughs in the late 1980s led to a re-evaluation of potential verification mechanisms for the BTWC. An important factor in this re-evaluation was the acceptance of on-site verification in arms control between the US and the USSR and the near completion of the Chemical Weapons Convention (CWC). Its verification provisions represented one of the most intrusive and rigorous regimes ever to be agreed. Coupled with this, the end of the Cold War, the Persian Gulf war and the suspicions of up to seventeen states possessing BW programmes, all added impetuous for active consideration of a verification regime for the BTWC. This was matched by increasing progress in biotechnology and the recognition of its applicability to weapons development and an increased aca-demic interest in the BTWC, which stemmed from fears that scientific break-throughs being achieved in the biological sciences would be turned to the advantage of the military. Thus, by the time of the Third Review Conference the discussion among the States Parties related to verification was not a the-oretical issue. In addition to the improvements in the CBMs it was agreed to assess the feasibility of verification. While this was a step forward it should be recognised that this action in itself was a compromise solution. In fact, as the former deputy-head of the US delegation noted in 1994:

> The issue of verification became the single most contentious question at the 1991 BWC Review Conference. The majority of States parties argued that they should incorporate verification measures into the BWC even if those measures were not completely effective since such measures would contribute to deter-ring BW proliferation. The United States, however, argued that the BWC was not verifiable and it had not identified a way to make it so. In simple terms the argument was between those who considered 'some verification was better than none' and the United States which argued that 'bad verification was worse than none.' (Moodie, 1994: 2)

Thus, the compromise solution by the States Parties was the creation of the Ad Hoc Group of verification experts, the so-called VEREX process, 'to identify and examine potential verification measures from a scientific and technical standpoint' (United Nations, 1992). The US problem with verification focused on technical issues, however it should be recognised that the fundamental difficulty was political rather than technical, in that the debate centred on whether verification would impede more the verifier or the violator (US Congress, 1993b: 75). VEREX reported in 1993 and a Special Conference adopting the report established another Ad Hoc Group, 'to consider appropriate measures, including possible verification measures, and draft proposals to strengthen the Convention, to be included, as appropriate, in a legally binding instrument, to be submitted for the consideration of States Parties' (United Nations, 1994). The Ad Hoc Group (AHG) has been meeting since 1995 in a series of sessions, which, by the end of 1999 totalled forty-four weeks of negotiation. It is now understood that the AHG has entered the final phase of its negotiations. Optimistically the Protocol could be completed by early 2000 and be adopted at a further Special Conference in late 2000, as per the European Union's position on the Protocol (United Nations, 1998). However, there were a sufficient number of 'false dawns' in the CWC negotiations to warrant some caution in this matter. There remain very real and protracted differences among the delegations. Pessimistically, it might be 2001 before the Protocol is completed, in time for the Fifth Review Conference of the BTWC, or, in a worst case scenario, even beyond this date.

The emerging structure and potential outcomes

Second-guessing the AHG is fraught with difficulties. Vogler and McGraw (chapter 8) notes in this volume that the Convention on Biodiversity required a 'control protocol' on transfer of living modified organisms. This Protocol, as he goes on to note, was derailed in February 1999 when most observers expected the Protocol to be completed. A similar event could befall the verification protocol for the BTWC. However, the basic parameters of the future verification regime are sufficiently well known from the detailed mandate established by the States Parties – other arms control agreements, such as the CWC – and open-sources covering the negotiations. Any stumbling block is likely to be of a political rather than a technical nature. The structure of the Protocol is likely to consist of (Pearson, 1998d: 191):

- mandatory declarations of facilities and activities most relevant to the BTWC, for example laboratories defined as Biosafety Level 4 and facilities working in biodefence activities;
- provision for visits to declared facilities, the main purpose of which will be to confirm the accuracy of declarations and/or clarify (mis)interpretation of declarations;

- provisions for the investigation and resolution of non-compliance concerns among States Parties;
- specific procedures for the investigation of alleged use of a biological weapon(s);
- measures to enhance peaceful co-operation among States Parties (Article X);
- improved national implementation of the Convention (Article IV);
- strengthened non-transfer procedures to prevent proliferation of BW (Article III).

One, but by no means the sole remaining area of contention revolves around a perceived dichotomy in the Convention; the requirement not to assist in the transfer of prohibited activities (Article III) and the requirement to assist in peaceful co-operation of activities relevant to the Convention (Article X). The other areas of contention have precedents in existing arms control agreements, and while it is not simply a case of 'one solution fits all' there is sufficient common-ground from which a solution could be derived from adapting and modifying existing practice. The control versus co-operation issue is unlikely to be resolved this way since the same problem in the CWC is deemed to have failed, as illustrated below. Thus, the AHG may need to consider new approaches.

The insertion of a co-operation procedure in the BTWC was not an uncanny piece of foresight in the early 1970s with an eye on the pending biotechnology revolution. Rather, it stemmed from current practice in arms control. Although all States Parties were willing to denounce BW, it had become practice in arms control and disarmament agreements to include safeguards against hampering the peaceful uses of technology, usually in the form of peaceful co-operation obligations which also acted as an incentive to join the treaty. In the area of nuclear weapons this can be seen in Article IV of the 1968 Nuclear Non-Proliferation Treaty. At this time, the late 1960s, there was widespread optimism about the utility of peaceful nuclear technology for the supply of cheap energy. In the 1993 CWC, Article XI of that agreement requires peaceful co-operation. However, as civilian nuclear technology has become less important over time, peaceful use of biotechnology has become more important since the entry into force of the BTWC. Subsequent review conferences of the BTWC have repeatedly flagged concern about the implementation of Article X and the increasing gap in biotechnology between the developed and the developing states (United Nations, 1986, 1992 1996; Sanhueza, 1996). The problem, in the opinion of a number of developing countries, is compounded by the developed states use of export controls on sensitive technologies to prevent proliferation of weapons of mass destruction. Coupled with this is the perception of the developing states that the agreement reached in the CWC, with respect to the same problem, that is, to prevent proliferation without hampering technological development, has failed to deliver a satisfactory solution. The chairman of the AHG noted:

> The solution which was agreed in respect of the CWC was an undertaking to revisit the need for export control regimes in a couple of years time. A similar solution seems unlikely to be acceptable to the developing countries in respect of the BWC. These countries consider that the undertakings under the CWC are too weak and that these undertakings are not being met, and consequently the developing countries are keen to see a true elimination of export control regimes in the verification protocol. However, several industrialised countries regard export control regimes both now and in the future as being a necessary element to meet their obligations under Article III of the BWC as an important ingredient of an integrated non-proliferation regime. (Tóth, 1997: 4)

These two issues, non-proliferation versus peaceful co-operation, are intrinsically linked. The seeming intransigence of the issue is based on a perception that one factor cannot be achieved without a detrimental impact on the other. That is to say, non-proliferation strategies for sensitive technologies in the form of export controls have a detrimental effect on peaceful co-operation in biotechnology. There is reason to reconsider this approach. An evaluation of export controls in the US, related to biological weapons, resulted in the conclusion by one analyst that:

> The conventional wisdoms about export controls on items and materials sensitive from the point of view of biological weapons have it wrong. They are not trade restrictions, but trade enablers. The trade that results is not controlled in a traditional sense, but licensed. The export control system is most accurately described as a system that licenses, channels and renders transparent trade in dual-use materials. (Roberts, 1998: 241)

One way forward might be linkage of security and environmental regimes (Pearson, 1997). The (draft) Biosafety Protocol's advanced informed agreement for the transfer and handling of modified organisms would offer one such opportunity (Pearson; 1998a). Such procedures are important for environmental safety considerations, but serve also a security function in their transparency of action: 'They share a common aim of ensuring transparency in the use and transfer of biological materials, thereby building confidence that such materials are not being misused' (Pearson, 1997: 372). Other 'linked' areas include laboratory safety, good manufacturing practice and disease surveillance in conjunction with a variety of other international organisations, such as WHO (Pearson, 1998a–c). Some States Parties in the negotiations have recently picked up the dual-purpose approach for the BTWC Protocol. In this example the Protocol visit regime, to ensure the accuracy of declarations, might also serve as a conduit for peaceful co-operation via assistance in areas such as good manufacturing practice or health and safety issues in laboratories (United Nations, 1999). Specific procedures could therefore be adopted in other areas to both strengthen and improve peaceful co-operation and enhance confidence through transparency.

If this duality was to be fully integrated as a central strategy in the Protocol, it represents one possible resolution to the contentious issue of control versus co-operation. In this respect the framework could be that existing export controls, for example those of the Australia Group, are brought into the Protocol to be enacted at the national level by all States Parties and administered by national authorities in conjunction with the future BW Organisation. The *quid pro quo* for the removal of *ad hoc* export controls among States Parties would be three-fold. First, the regulation of trade in those technologies most relevant to the Convention, such as fermentors. Second, the application of licensing procedures to ensure end-use of equipment. Third, the prohibition of trade in such items with non-States Parties. The Federation of American Scientists (1995) has mooted similar proposals. This dual-purpose approach offers a feasible way forward for the AHG to resolve its difficulties in the control versus co-operation issue. 'The challenge is to identify measures that will enhance the implementation of Article X whilst *at the same time* contributing to the building of confidence that States Parties are compliant with the convention, and avoiding duplication with activities in other fora' (Pearson, 1998d: 196).

Addressing the BW problem will not produce results overnight. After the recognition of the Soviet and Iraqi BW programmes confidence in the BTWC needs to be restored, but confidence and trust take time to build. While strengthening the Convention is essential it will not immediately remove all doubts about the proliferation of BW capabilities nor, of itself, prevent the future use of biological weapons. As the Chairman of the negotiations noted, 'the verification protocol will not and cannot be the answer to all such challenges' (Tóth, 1997: 2) and consideration needs to be given to other strategies and new international law. One recent proposal would be to criminalise the actual use of biological weapons (Anonymous, 1996; Crawford *et al.*, 1998). The proponents of this approach note that neither the BTWC nor the CWC makes the prohibited acts an international crime. Nor do other international treaties or conventions. They go on to state that:

> What is needed is a new treaty, one that defines specific acts involving biological or chemical weapons as international crimes, like piracy or aircraft hijacking, obliging states either to prosecute or extradite offenders who are present in their territory. Treaties defining international crimes are based on the concept that certain crimes are particularly dangerous or abhorrent to all and that states therefore have the right and the responsibility to combat them. Certainly in this category, threatening to the community of nations and to present and future generations, are crimes involving the weaponization of disease or poison and the hostile exploitation of biotechnology. (Crawford *et al.*, 1998: 1)

Such a treaty would, the proponents note, strengthen the norm against chemical and biological weapons, enhance the deterrence against potential offenders and facilitate international co-operation to suppress prohibited activities (Crawford *et al.*, 1998: 2). Interestingly, this returns to an idea expressed in 1969 during US government hearings on chemical and biological weapons. Of the experts who gave testimony one noted that the impracticality of verification might be addressed by a more general-purpose compliance mechanism that would strengthen the norm against biological weapons by effectively making the use of biological weapons a crime against humanity (US Congress, 1970: 114–15).

Conclusion

The problems facing the international community with respect to BW are not new. Both the Geneva Protocol and the BTWC sought to address the concerns of the international community by prohibiting biological weapons and enshrining such agreements in international law. Events since 1975 have demonstrated that the BTWC failed to produce a sufficient amount of confidence in the regime to consider the problem resolved and developments in biotechnology since that time have compounded the issue. As illustrated in chapter 2 (Manning) of this volume biotechnology uses expertise from numerous disciplines. Equally, the US Congress has noted that it is all but impossible to control the further dissemination of weapons-relevant information from biotechnology. Compounding this further is the observation by Dando (chapter 10, this volume) that advances in biotechnology will produce capabilities that could be turned for malign purposes. Thus, the complexity of the problem is: the application of a technology which relies on an interdisciplinary collection of knowledge that is already widespread, and likely to become increasingly so, will, as an offshoot, produce capabilities that could be used for biological weapons. If we understand this to be the problem, a simple solution, which is satisfactory to all the parties concerned with preventing biological warfare, is an extremely unlikely outcome. Second, it does not bode well for strategies that are not universal. This suggests that an overlapping and integrated strategy is required.

Should this not be the case, and the parties negotiating the Protocol fail to agree upon a comprehensive strategy to the satisfaction of all, then the series of individual strategies based upon export controls, unilateral sanctions and pre-emptive military strikes against suspected proliferators of biological weapons might prevail in the next few decades. Such an outcome would be wholly unsatisfactory. Many analysts consider the time is right for a comprehensive BW regime, but there is a very real danger that the technology could outpace the (future) control regime if agreement is not reached soon. This problem, technology outpacing regulation, is not solely confined to biotechnology, but is apparent also in the other technological driver of this decade: information technology. In a recent article the Vice President of IBM noted that the export control regime in the USA that limited the sale of 'super-computers' to other nations was, in terms of the threshold of computer power, behind the advance in computer technology. Thus, 'computer speed and performance are increasing so rapidly that they have rendered meaningless current US policies concerned with safeguarding national interests and security' (Caine, 1999). He went on to illustrate that dual Pentium III processors in business PCs (as of April 1999) already performed in the range restricted by US export controls and that any belief that raw capability could be controlled was anachronistic. He concluded that: 'We need to develop a fundamental philosophy on export controls that is in tune with the technological, economic and security realities of the future' (Caine, 1999). This could easily have been written for biotechnology and its application for weapons purposes. Whilst we should not lose sight of the fact that the BTWC and its future Protocol relate to disarmament, the biotechnology revolution and the dual-purpose nature of the knowledge and technology required for BW means that the control regime has implications and responsibilities beyond the confines of national security. Thus, an integrated strategy will be required if the regime is to be successful and this makes biotechnology and its future applications different from the chemical industry and the nuclear industry.

Notes

1 *Convention on the Prohibition of the Development, Production and Stockpiling of Bacteriological (Biological) and Toxin Weapons and on their Destruction.* Signed on 10 April 1972. Entered into Force 26 March 1975. Depository Governments, United Kingdom, United States of America, Union of Soviet Socialist Republics. The Convention is referred to as either the Biological and Toxin Weapons Convention (BTWC) or the Biological Weapons Convention (BWC), whereby toxins are included in the generic term 'biological weapons' in the latter example.

2 Once the negotiations for a treaty or convention are finalised the treaty is opened for signature. A country is a 'State Party' to an international treaty or convention when it has both signed the Convention and ratified it through its own domestic legislature procedures. In the case of the UK this would entail signing the Convention and then ratifying it through the Houses of Parliament. Countries that only sign the Convention are known as *Signatory States*. Once an agreed number of countries have signed and ratified the treaty the agreement *enters into force*. States Parties are legally bound by the obligations of a treaty once it has entered into force. Signatory nations are not.

13

Responses to security threats posed by biological weapons

Joanna Spear

Following on the chapters by Dando and Whitby on the new types of biological weapons (BW) we are likely to face in the future, this chapter discusses the issue of responding to the threats posed by biological weapons. It is the central contention of this chapter that the biotechnology revolution is having a profound effect on the degree of threat posed by biological weapons (BW). This means the task of dealing with the problem of BW is becoming all the more urgent and more difficult.

The chapter concentrates on five aspects of the issue which form its analytical framework. The first section examines the problems with the *denial* strategies which are employed by the international community to choke off the biological weapons problem at source. In contrast to the chapter by Littlewood, this section discusses the shortcomings of the Biological and Toxin Weapons Convention (BTWC) and other elements of a denial strategy. The second section focuses on attempts to *deter* biological weapons deployment and employment, advocated by those who claim that attempts at denial have either failed, or are not a sufficient guarantee of security. Section three focuses on the problem of *detection*. This is a crucial problem in two senses, first confirming whether an outbreak of a disease is a natural occurrence or the result of biological weapons use, and, second, whether detection can be done in a timely manner. The fourth section looks at the question of *defence* measures. The final section examines the problem of *destruction* of biological weapons. This could occur as a forceful response to biological weapons use, from a successful pre-emptive action (of the sort we saw in Iraq with the United Nations Special Commission (UNSCOM) programme), or as a result of a verified biological weapons disarmament regime.

It is essential to recognise that there are two levels of threat to be considered when dealing with biological weapons. First, there is state use of biological weapons (which would also encompass threat or use by a state-sponsored terrorist group). The state is the traditional source of BW threats, but the changes in biotechnology are increasing the ability of rogue states to pose security threats through BW. Secondly, there is the less familiar issue of BW threats from non-state actors. As Falkenrath has argued, the threat from such terrorists using BW is growing because these weapons are increasingly easy to manufacture and disseminate, and this is coupled with a decrease in the moral and practical restraints against causing mass casualties. (Falkenrath, 1998). The wide dissemination of the fruits of the biotechnology revolution is having the effect of making the threats of BW use from both the state and non-state level more serious and credible.

The reason that the issue of BW threats has to be re-visited is because many of the factors which in the past inhibited the development, manufacture and dissemination of BW can be overcome thanks to the biotechnology revolution. BW production and use was previously inhibited by the significant problems which surrounded its manufacture (the need to be able to mass produce the agent; the need for a short and predictable incubation period), its dissemination (the need to be able to disseminate the agent effectively without impairing its biological properties; the need for it to be effective in low doses; and the need for the agent's effects to be consistent) and its consequences (the need for the target population to be vulnerable to the disease, with no treatments available in a timely fashion; the lack of discrimination of BW and the consequent fear of loss of control) (Dando, 1994). The biotechnology revolution is facilitating the development of stable and predictable BW, it is allowing the development of 'novel' organisms against which there is no immunity nor readily available antidotes and it is facilitating mass production, storage and dissemination (Bartfai *et al.*, 1993: 301–4). For these reasons the issue of the security threats posed by biological weapons merits renewed attention.

This discussion also highlights some of the logistical issues and dilemmas associated with BW which are of relevance when considering attempts to forestall their production: the fact that the pathogens are likely to be small, stable and easily concealed; and the fact that they can be mass-produced in non-dedicated facilities. It also usefully points to some of the key elements in combating the threats posed by biological weapons: the need for immunisation, treatment, and various forms of protection against the diseases and toxins which could potentially be disseminated.

Concerns have been expressed that biological weapons might be a 'poor man's nuclear bomb' because they are the only weapon to match nuclear weapons in terms of potential for human destruction. In some ways though, BW are *more* problematic than nuclear weapons for the international community. Greater hurdles have to be overcome in order to acquire the materials, technologies and facilities to enable the development of a nuclear weapon. Moreover, significant technological expertise is required to weaponise a nuclear device. By comparison, BW are relative simple to develop, materials are easy to obtain legitimately and there are low-technology ways in which they can be delivered, for example, by crop sprayers or human delivery. More sophisticated delivery systems (for example, aerosols, spray tanks, bombs and missiles) do raise hurdles to BW deployment and use. Indeed, weaponisation is the most difficult part of BW production. However, even this problem is diminished in the era of massive advances in biotechnology.

One of the key dilemmas attached to the issue of biological weapons is the inherent dual-use capabilities of many of the components of a BW. They can literally be used for good or evil. The dual-use dilemma is heightened by developments in biotechnology: the negative implications of many biotechnology innovations, genetic research, etc., need to be considered. Genetic research certainly advances the cause of medical science, but such work could be perverted to enable the targeting of BW against particular ethnic groups (Bartfai *et al.*, 1993: 304). Thus, for example, the Human Genome Project, which has great potential for enabling us to eliminate certain genetic diseases, also provides knowledge which could be used to develop ethnically targeted BW (British Medical Association, 1999: 60). Although this may seem far fetched, ethnically targeted weapons were under development in apartheid South Africa in the 1980s and early 1990s (Beresford, 1999: 2). The inherent dual-use capabilities of the components of biological weapons, biotechnology techniques and knowledge are the complicating factors which inform this chapter.

Deny

Attempts to deny states and sub-state groups BW capability work on several different levels and employ several different means, some inherently peaceful, some involving the use of force. Amongst the peaceful denial strategies are diplomacy, international law and building regimes with strong norms. Later in the chapter the destruction of BW facilities will be discussed, but this should be borne in mind as an element of a forceful denial strategy. This section will look at the (peaceful) denial strategy of regime formation and the supporting strategy of forming a strong norm against BW deployment and use. There is an increased sense of urgency about the search for an effective, peaceful denial strategy. This is necessary because, thanks to the biotechnology revolution, BW are becoming more feasible and usable by a variety of actors. The race is on to impose an effective, verifiable regime before BW are used.

Littlewood's chapter discusses the Biological and Toxin Weapons Convention (BTWC) in some detail and sets out the efforts underway to strengthen this centerpiece of the BW regime. As the efforts to strengthen it indicate, the BTWC is currently very weak. A number of crucial problems with it can be identified.

Weaknesses of the BTWC
One of the crucial problems of the existing BTWC is that it primarily

addresses the issue of *state* production of BW. However, there is increasing international concern about the activities of *sub-state* groups who regard BW as usable instruments of power. The Aum Shinrikyo sect's use of Sarin gas in the Tokyo subway system in 1995 was particularly enervating. This was because the sect's activities had not been considered worthy of particular attention by the Japanese authorities. Despite the assumption that they were not dangerous, they actually caused both death and mass panic, and subsequent investigations into the sect revealed a systematic attempt to obtain a BW capability. For example, the group unsuccessfully tried to use botulinum toxin and anthrax (Carus, 1997: 3, table 1). It was discovered that the sect's members had gone to work in Africa as 'volunteers' following an ebola outbreak there! (Cole, 1997: 3). Importantly, the group was actually attempting to develop a method of aerosol dissemination – a sophisticated weaponisation which would produce higher fatalities and be more problematic for authorities to deal with. The activities of the Aum highlight a key problem for the BTWC, even a strengthened regime would have minimal effect on the activities of such a sub-state terrorist group.

Another crucial weakness of the existing BTWC concerns verification and the extent to which the Convention can be relied upon to ensure that states do not develop BW capabilities. A connected problem is that the development of defences against BW is *intimately* connected with the development of offensive BW capabilities. As Geissler noted:

> the blurring of defensive and offensive programmes is not limited to research, but extends to development, testing, production and training. Although it may be possible to use the scale of the project as an indicator of the intent at the development stage and beyond, there is a large grey area where offensive and defensive needs will be indistinguishable. (Geissler, 1986: 67)

This therefore produces tremendous verification dilemmas when trying to contain the development of offensive BW capabilities by threatening states, who claim to be merely exercising their right under the BTWC to be able to defend themselves. Many states who are party to the Convention have ostensively defensive BW programmes. As discussed by Littlewood, various initiatives are underway to address the issue of providing adequate surety that the regime is effective and no states parties are evading it. However, these initiatives have yet to bear fruit. There are therefore incentives for states to use the BTWC as a cover for developing their BW capabilities. More than that, the lack of verification in the BTWC might be a causal factor in leading a state to develop an offensive BW capability, due to a 'security dilemma' exacerbated by the lack of verification in the regime.

Amongst signatories to the BTWC there remain questions over whether they have actually complied with the terms of the convention. For example, in 1998 there were allegations from a defector who, until 1992 was a senior official in the Soviet biological warfare programme (which President

Gorbachev was supposed to have cancelled in 1990). Dr Kanatjan Alibekov alleges that work has continued on offensive BW capabilities under the guise of defensive research. This seems to have been spurred by the belief amongst his former colleagues that the US was also secretly continuing to develop its BW capabilities (Fairhall, 1998: 12). Dr Alibekov has also suggested that information from the Soviet/Russian programme was passed on to other states, including Serbia, which has been interested in BW for years (BBC Today, 1999).

The 1991 Gulf War against Iraq brought to the fore concerns about the reliability of the BTWC as a guarantor of a BW-free world. The allied forces were aware of the potential for Iraqi use of both chemical and biological weapons against them. The subsequent revelations about the Iraqi BW development programme and the discovery of stores of biological agents have done nothing to strengthen belief in the BTWC, which Iraq had signed in 1972 (though it did not ratify the convention until 1991) (Whitby and Rogers, 1997).

Moreover, the most intrusive investigations into a BW capability to date, the UNSCOM inspections in Iraq, have not increased faith in the international community's ability to verify whether a state is developing BW. As a former UNSCOM inspector concluded:

> The UNSCOM experience demonstrates very clearly how difficult it is to achieve biological weapons arms control. For instance, despite having powers in Iraq that no other international compliance could ever hope to have, and being able to call on help from the intelligence services of the most powerful nations in the world, UNSCOM has not been able to clarify Iraq's former BW program to the point where it feels it has a sufficient basis to commence OMV [ongoing monitoring and verification] in the biological field ... The disquieting inference I draw ... is that at present methods and approaches to biological arms control are inadequate and require reassessment. (Zilinskas, 1995: 235)

These concerns have become even more pronounced with the subsequent hiatus in intrusive inspections which resulted from *Operation Desert Fox* mounted in December 1998.

A crucial weakness of the current BTWC is that it has no intelligence gathering facilities of its own. The regime is therefore reliant upon the information provided by states parties, although this carries the danger of disinformation being provided for political purposes. Some intelligence can be gathered through on-site inspections, but given the short lead times of BW programmes – ten billion infectious doses of anthrax spores could be developed within a week using a 100-litre fermenter (Norton-Taylor, 1999: 8) – even inspections are of limited value.

In sum, the current state of the BTWC offers incentives and opportunities for states concerned about what other states parties (and non-parties and sub-state groups) are doing to push ahead with (ostensively) defensive BW developments in ways which may lead to destabilising security dilemmas.

More importantly, the improvements to the BTWC currently being painstakingly negotiated are not likely to solve all the problems laid out above, particularly the lack of attention to sub-state threats of BW use.

The BW use taboo

One of the supports to the BTWC and other efforts to prevent the development, deployment and use of BW is the widespread psychological revulsion against BW. This has created what some have described as a taboo against BW. Interestingly, according to Falkenrath, even amongst non-state groups there are qualms about nuclear, chemical and biological (NBC) use and 'This norm against NBC use is probably strongest in the case of biological weapons' (Falkenrath, 1998: 53).

However, there are three issues worth noting here. First, the taboo must be of fairly recent origin as many states had offensive BW programmes prior to the BTWC. Second, the idea of a taboo against BW may be a Western concept and actually have no sway outside of the West. Thus, the taboos against BW may be weakest in the areas of the world where we would most want them to operate, that is, where there are rogue states or sub-state groups. For example, in 1996 the US Central Intelligence Agency (CIA) estimated that around ten countries in the Middle East and South Asia were developing or had chemical and biological weapons (CBW) (CIA, 1996: 1). Third, even within the western countries where the taboo is supposed to operate, there have been an increasing number of incidents of threatened or attempted BW deployment by sub-state dissident groups. Despite noting the past norm against BW use amongst non-state groups, the thrust of Falkenrath's work is to discuss the erosion of that taboo (Falkenrath, 1998). Despite the apparent erosion of the taboo against BW, at present biological weapons have not been assimilated into the arsenals of most states – thanks to the BTWC – which potentially facilitates their effective banning. Time is therefore of the essence, to formulate an effective, verifiable, regime for eliminating BW before the taboo is completely eroded. This also suggests a strategy for strengthening the BTWC and the norms of non-use: target resources at the states and militaries where it appears the taboos have not been accepted. A campaign of public 'demonisation' of the weapons may also serve to shore up the norm of non-use amongst non-state groups – at least in the short term.

As part of a denial strategy, greater monitoring of the supply of active diseases to states and groups purportedly working on cures for diseases would seem to be necessitated following revelations about both Iraqi and Aum successes in getting active diseases from the Center for Disease Prevention in Atlanta, and the near-successful attempt by a white supremacist to obtain bubonic plague bacteria from the Maryland firm ATCC (Cole, 1997: 156–7). Given the inherent dual-use capabilities of many biotechnologies, denial of access to organisms, etc. is a particularly difficult strategy. This suggests that denial strategies should be primarily targeted at the weaponisation stage, rather than at the precursor biotechnologies. However, concentrating on the weaponisation process would not help when faced with states or groups employing very simple dissemination techniques.

It is clear from this brief review that peaceful denial strategies have not been completely successful and are under increasing pressure. Time is running out for peaceful denial strategies because of the increased usability of BW thanks to the biotechnology revolution. Efforts are underway to shore them up, but many are now putting greater emphasis on other responses to the BW problem.

Deter

According to Kathleen C. Bailey, time spent trying to shore up the BTWC-based denial regime is time wasted. She calculates that the combination of the verification problems inherent to the regime and the costs to the biotechnology and pharmaceuticals industries of verification measures makes arms control a waste of time. Rather she advocates that efforts should be devoted to 'deterrence to forestall use or threat of use of BW' (Bailey, 1995).

In effect this was in operation during the 1991 Gulf War when the US-led coalition left open the possibility of responding to Iraqi WMD use in kind (Freedman and Karsh, 1993: 255). In actual fact, the option had been ruled out in US planning, but the threat was thought to be potent. Deterrence was not generated by the US alone. The threat of Israeli WMD retaliation was a serious consideration. The subsequent discovery of the advanced state of Iraqi WMD capabilities, particularly its chemical and biological programmes indicates that a form of deterrence did operate as Iraq did not use the CBW artillery shells it had prepared. It has been suggested that Saddam Hussein calculated that he could survive as long as he did not use chemical or biological weapons (Freedman and Karsh, 1993: 434).

It is worth noting that the argument for deterrence against BW is actually an argument which has been used to justify creating a biological weapons capability. For example, allegations were made before the Truth Commission in South Africa about the apartheid state's efforts to create a biological weapon which would only affect blacks. The Managing Director Daan Goosen of the Roodeplaat Research Laboratory (the front company for the military), 'justified the project by comparing it to the nuclear arms race, saying the intention was not to use the technology, but to secure peace through fear' (Beresford, 1998: 15). This line of reasoning also further blurs the boundary between an offensive and defensive BW programme.

Detect

Detection needs to occur at several different points in the BW process. First, there is the issue of detection of clandestine BW production facilities. Second, there is detection of weaponisation of BW and movement to suitable launch areas. Finally there is the issue of detection of BW release into the environment (land, water or air). At each stage in the process, there are significant problems in detecting biological weapons.

What forms of detection are required will vary depending upon who the suspected BW proliferant is. For example, a state thought to be producing BW could rule out all intrusive detection techniques if it was outside of the BTWC. A member of a strengthened BTWC would be obliged to allow inspections and other detection techniques. However, a suspected bio-terrorist group (where there would probably be no sophisticated weaponisation) would present a qualitatively different challenge, particularly if the group was able to shelter within a state not a member of the BTWC.

Clandestine facilities

As was noted above, BW production facilities may be little different from civil biotechnology production facilities (except perhaps in scale). Amongst the facilities necessary for large-scale propagation of agents used in BW production are fermenters, incubator shakers, air filtration systems, scrubbers and cold storage facilities. A number of protective technologies are also required (although Iraq seemed to have few of these). This means that an effective global monitoring system which combined information about biotechnology equipment purchases might play a useful role in detection. Thus, the Wassenaar Agreement has a potential role to play here. Various methods of detection of clandestine BW facilities have been identified:

- comparison of declared civil production facilities with actual facilities;
- on-site inspections;
- sampling and identification;
- inventory control of dual-use equipment;
- on-site continuous monitoring;
- aerial surveillance;
- epidemiological surveillance (Zilinskas, 1995: 233–5).

As Littlewood has noted (this volume), some of these techniques are being considered for inclusion into a verification scheme for the BTWC.

Many of these techniques were employed by the UNSCOM inspection teams working in Iraq. The UNSCOM teams had unprecedented advantages in their work to verify Iraq's claims about its BW activities. Specifically, they enjoyed a level of access which is unlikely to be replicated and also received some state co-operation with their verification work (particularly after September 1993). Even with these advantages, UNSCOM has experienced real difficulties in trying to piece together the story of Iraq's BW programme (UNSCOM, 1998). One point worth noting about the UNSCOM inspections has been their heavy reliance upon information provided by defectors. The UNSCOM attempts to uncover for themselves the full scope of Iraqi WMD developments have not been successful. This is a salutary lesson in many ways, showing that verification remains a key problem. As Zilinskas concluded after participating in the work of UNSCOM:

> Given the ease of hiding small but significant quantities of potential BW agents, barring a fortuitous accident there is no possibility for an outsider to ascertain if seed cultures of pathogens that potentially could be developed into biological weapons are present somewhere in the country. (Zilinskas, 1995: 233)

The problem of detecting clandestine BW facilities of sub-state groups may be somewhat easier if the host state is co-operative. However, if that is not the case, then problems are severe. These problems necessitate a lot of emphasis on good intelligence, particularly that from spies and defectors (see below).

Weaponisation and movement

Remote sensing might have some role to play here, but detection of weaponisation and movement is made more difficult by the compactness of most biological weapons (in comparison to nuclear weapons). Detecting redeployment would require a significant amount of information about what 'normal' movements were. We also need to bear in mind the way that India was able to avoid satellite detection of its initial nuclear test by careful calculation of when there was no satellite coverage of the test areas. Additionally, the intensive remote sensing necessary for detection of BW movement would involve heavy use of satellite time – an investment only likely to be made in situations of acute tension, when threat of BW use is imminent. Moreover, this does not help with the issue of the intentions behind or targets for pre-emptive strikes with BW systems.

Detection of release

Detection might involve recognition of use in the field, laboratory confirmation of the pathogen(s) detected and a suitable system for informing relevant authorities of the problem. There is a need for timely detection and analysis and much effort is being put into improving the ability to detect a wide range of BW and decreasing the time between detection and analysis. This is where the biotechnology revolution is playing a positive role, with the use of modern molecular genetic methods to diagnose the presence of unusual pathogens. The United States Defense Advanced Research Projects Agency (DARPA) recently put out a call for proposals for 'revolutionary' systems able to detect a wide range of biological agents, including bio-engineered viruses and bacteria. The project is primarily interested in the military utility of these systems, but they should also have positive spill-overs for verification of compliance with the BTWC (*Jane's Defence Weekly*, 1998: 8). Britain has recently deployed a prototype BW detection system with its troops in Kuwait (Norton-Taylor, 1999: 8).

An important aspect of detection is the need to determine whether an outbreak of a disease was naturally occurring or a consequence of deliberate spreading of a biological pathogen. This is important for reasons of retaliation, defence against further attacks and the use of international law against the perpetrators. However, it is also important for obtaining information relevant to an antidote (particularly if the BW was a hybrid for which the attacked state did not possess knowledge of).

Determining whether a pathogen is naturally occurring or man made is not as easy as may be thought. This is particularly the case because changing weather patterns are heralding the return of forms of disease which were hitherto thought to be extinct or severely restricted, and would therefore have been attributed to man rather than to nature. The effects of weather systems such as El Nino and general climate change as a result of the effects of global warming means that, 'In effect, humans could inadvertently unleash biological warfare on themselves' (Pearce, 1998: 2). Pathogens are also capable of mutating and adapting. For example, an outbreak of Rift Valley Fever, which traditionally affects cattle, seems to have jumped the species barrier and subsequently killed hundreds of people in the north east of Kenya (Pearce, 1998: 2). Moreover, there is also a class of what Morse has termed 'emerging viruses' which are new or rapidly expanding viruses in the human population. Examples include, HIV, Ebola virus, denguee virus and hanta virus (Morse, 1994: 168–9, cited in Cole, 1997: 220). When first encountered, it might be assumed that these naturally emerging viruses are actually a form of BW.

Another situation can add to the confusion over the sources of a biological pathogen. It is the case, particularly in countries with advanced medical services, that drug resistant bacteria are emerging due to the overuse of antibiotics. Amongst the micro-organisms with drug resistance are malaria and tuberculosis (Bartfai *et al.*, 1993: 297). This raises the spectre of new strains of a disease occurring naturally, through this mutation process, but being difficult to treat because of their in-built resistance to the classic medical treatments.

The issue of natural occurrence versus malign forces has come up at several points in the past. For example, during the Second World War an outbreak of Spanish Influenza in Britain was blamed on Nazi biological weapons. The most recent case where there were questions over whether a disease was natural or a consequence of BW use occurred in 1997 at a meeting of the BTWC. At the meeting the Cuban government accused the US of spraying Cuban potato fields with a crop-eating pest in October 1996. Cuba was the first state to invoke the 1991 provision of the BTWC which allows a country which believes it has experienced biological attack to request a consultative meeting so that the issue can be investigated. Cuba's claims could not be proved, as the pest, *thrips palmi* can be naturally transmitted by winds (Reuters, 1997: 8).

As well as adding to the problems posed by BW, developments in biotechnology will have some positive effects on detection. Biotechnology will provide a means to speedily establish the type of pathogen, to decide whether it was a BW or natural outbreak, and – if it is not a known disease strain – most importantly, provide the means to trace who developed it. This is because 'novel' organisms (from which a designer biological weapon would be made) leave a distinctive 'tag' or 'fingerprint' which can be traced. Indeed, designer weapons may be easier to trace back to the perpetrators than BW using existing disease strains.

Intelligence

In all areas of detection, intelligence plays an important role in directing the attention of authorities to potential BW threats – be they states or sub-state groups. In addition to the intelligence shortcomings of the BTWC noted above, many nations are now realising the inadequacy of their intelligence for BW threats. Different forms of intelligence are suited to different types of problems. Although signals intelligence and electronic intelligence (SIGINT and ELINT) may be of some utility in detecting BW threats, it is thought that the best form of intelligence is human intelligence (HUMINT) given how easily concealed BW are. HUMINT is also the most difficult to obtain as both state or sub-state BW programmes require limited manpower. This means that the opportunities for HUMINT are limited.

The US General Accounting Office (GAO) released a report in December 1997 which highlighted the inadequacy of US government co-ordination of efforts to counter terrorist threats (Miller and Broad, 1998: 13). One of the strengths of the US is its SIGINT and ELINT; however, it has traditionally been weaker in the field of HUMINT and, during the 1991 Gulf War, relied heavily on human intelligence provided by East European intelligence agencies. The US had, until the early 1990s, not devoted much attention to American right-wing terrorist threats, having been more concerned about externally based terrorist groups. Nevertheless, US HUMINT capabilities (combined with tip-offs) did lead to the charging of a man connected to the 'Aryan Nation' (a white supremacist group) suspected of planning to spread anthrax on the New York underground system (Reed, 1998a: 1; Reed 1998b: 15).

Defend

Defence against BW could be merely passive or might involve offensive operations. In either case, one of the key requirements is accurate information. As discussed above, there is a requirement to improve the intelligence available on potential BW threats from states or bio-terrorists. Defence primarily needs to be considered at two levels, tactical and strategic. To deal first with the tactical level, it had been thought in the late 1950s that there could be no tactical application of BW (Balmer, 1997: 131). This was because of the delayed effects of the majority of BW and the inability of the perpetrator to precisely control the biological weapon (for example, a disease might spread beyond opposing forces to one's own forces). However, some of the developments in biotechnology are nullifying these problems. For example, efficient biotechnology delivery systems may mean that a particular disease acts more quickly and a BW could potentially be pre-programmed to 'terminate' itself after a certain number of replications. These developments could presage a new tactical utility for BW. This means considering the issue of defending troops in the field from various types of BW.

It has long been assumed that BW will be of maximum utility as a strategic weapon, striking at the heart of the state. Defence of civilian populations against BW is difficult to orchestrate. It may be possible to provide some protection through a ballistic missile defence system (BMD) which could destroy missiles carrying BW, but other methods of dissemination would be virtually impossible to defend against, for example, aerosol dissemination through the heating system in a skyscraper. In that situation, the best that can be achieved may be rapid response to BW use.

Another problem for defence against BW is posed by the threat of bioterrorism. The nefarious activities of the Aum Shinrikyo and rising concerns about terrorism generally has led to a new focus on sub-state BW threats. The rising interest in this aspect of the problem is reflected in international diplomacy.

Several categories of defence can be identified, including protection of humans, animals and crops. Some of the defences required for each of these are already available (though in small quantities) because of the need to combat normal disease outbreaks. Moreover, research into various pathogens, designed to increase crop production, protect domestic livestock and develop responses to human diseases, is obviously relevant here. Once again, developments in biotechnology can play a positive role (as well as a negative one). For example, advances in biotechnology and genetic engineering techniques are leading to improvements in immunology. 'Several immune system stimulants (colony-stimulating factors and cytokines) are now available as recombinant proteins. The increased understanding of the mechanisms of immune responses also helps in designing drugs and therapies to fight infection, and enhancing reactions to toxic substances' (Bartfai *et al.*, 1993: 299).

The type of BW opponent will determine the level at which defensive efforts have to be targeted. When dealing with a threat of BW use by a state, there needs to be (tactical) protection of troops in theatre and key transport nodes and if the state has ballistic missile capabilities, possibly also (strategic) protection of civilian populations. When dealing with sub-state groups the most likely threat is the use of BW against the civilian population using simple dissemination techniques such as crop sprayers, contaminating water supplies, etc. The ability to protect humans involves among other elements:

- vaccines, antibiotics or other prophylactic measures;
- buildings, protective clothing or other physical barriers;
- therapeutic measures;
- well-trained personnel for detection and protection;
- decontamination procedures. (Geissler, 1986: 67)

Protection of civilian populations is the aspect of the problem currently receiving a lot of attention, particularly in the United States. This is in part because in the wake of the 1991 Gulf War troop protection was increased significant. It was only in the mid to late 1990s that protection of civilian populations became a priority. This change was a consequence of both the activities of the Japanese Aum sect, the increased militancy of various white supremacist and militia groups, and the emergence of various religious cults with apparently murderous intent.

International activities

The British, Canadians and Americans have long been working together on the problem of defence against CBW. The 1980 Memorandum of Understanding (MoU) called for the maximum integration possible of the defence programmes of the three countries (Carter and Pearson, 1996: 95). This element of international collaboration, has come to complement the longer-running efforts of the Tripartite Technical Co-operation programme (TTCP) by focusing on short-term priority projects as opposed to the TTCP's focus on long-term research collaboration.

There has also been increasing attention to the CBW problem within the North Atlantic Treaty Organisation (NATO). The US has been pushing hard for the issue of defence against NBC threats to be taken seriously. Despite some objections to the US evangelism on the issue, the European states have agreed to NATO's adoption of a form of Counterproliferation (Carter and Omand, 1996). As a measure of concern about CBW, NATO's April 1999 meeting in Washington DC planned to address these proliferation concerns. As it was, the war with Serbia seems to have preoccupied the meeting.

At 1996 the Group of Seven advanced economies (G7) summit in Lyon, the Communique on Terrorism issued at the end of the meeting explicitly acknowledged the threat which would be posed by terrorists with a BW capability. Together these activities show the high profile the problem has in the advanced economies of the developed world – the most likely targets of bio-terrorism.

National activities

The US has taken the lead in tackling the issue of bio-terrorism, in part because of the international threats it faces, but also because of the activities of various extremist sub-state groups in the US itself. Thus, the US Counter-proliferation initiative explicitly discusses the problem of responses to BW terrorism. Indeed, it has been reported that of all the threats the US faces within the spectrum of WMD, the one of greatest concern to the Pentagon is BW use (Unattributable briefing, 1996). In 1996 Congress passed the 'Defense Against Weapons of Mass Destruction Act'. The Act, a direct response to the problems exposed by the activities of the Aum Shinrikyo in Japan, directs the President to enhance federal capabilities for dealing with WMD terrorist threats and to assist state and local governments in develop-ing their response capabilities. The DoD has a major role in supporting the initiative, along with the FBI and the Federal Emergency Management Agency (FEMA) (Carus, 1997: 4). The FY 1997 budget included legislation establishing the federal inter-agency programme to assist major cities prepare a response to CBW terrorism. As part of this planning, the Pentagon estab-lished a Chemical and Biological Quick Response Force (QRF) to assist twenty-seven major metropolitan areas likely to be subject to CBW attack by terrorists. Local emergency personnel in each city are being trained to respond to these forms of terrorism (Evers, 1997: 6).

Recent 'war games' carried out by the United States Government illus-trated the problems which would arise in trying to cope with a BW terror-ist attack. The scenario which was 'gamed' involved a virus being spread along the Mexican–American border. Amongst the problems which were highlighted were those of running out of supplies of antibiotics and vac-cines (presuming that one existed); getting trained, immunised medical staff into the affected areas; and coping with the panicking population. Altogether the game highlighted the fact that, given current preparations, the US authorities would be quickly overwhelmed by the disaster caused by terrorist BW use (Miller and Broad, 1998: 13). As a result of this experi-ence, the US is now planning to produce a nationwide plan to address the defence problems which the game exposed. Falkenrath *et al.* make a num-ber of suggestions for how the US might prepare to defend against terrorist NBC attack (1998: 261–336).

Destruction

The issue of destruction of BW is relevant to several different areas of the BW issue. First, it is necessary to consider whether there can be forceful pre-emptive destruction of BW. Second there is the question of BW destruction in the context of a victory in war, the best example of this being the activi-ties of the UNSCOM teams in Iraq after the 1991 Gulf War. Finally, there is the question of BW destruction in the context of a verifiable BW disarma-ment regime. In each of these contexts though, the dilemmas remain the same: Can BW be disposed of safely? Will all BW be destroyed? Will there be any lasting environmental damage from BW destruction?

Forceful destruction

We are already seeing pre-emptive attempts to deal with potential Chemical Weapon development facilities (US cruise missile attacks in Sudan and Afghanistan in August 1998) and there may be similar action against suspected BW production facilities in the future. The physical destruction of potential BW development facilities raises dilemmas in itself: Will all BW be destroyed? How long will this delay BW production? Will there be lasting physiological and environmental damage caused by forceful destruction?

Amongst the possible techniques for forceful destruction of BW capabilities are sabotage by special forces groups with BW training, air strikes either by conventionally armed manned bombers or cruise missiles (the former packing more of a punch and able to do effective reconnaissance, but more vulnerable to enemy defences) or, in a critical situation, the use of nuclear warheads.

A dilemma of forceful destruction is how to effectively destroy all BW without disseminating the materials across lands and possibly population centres (likely if the state or sub-state group has attempted to protect the BW facility by situating it in a urban area). Depending upon the particular pathogens involved, effective destruction may have to involve the use of ultra-high temperatures, for example, through the use of fuel air explosives. Some infective pathogens are almost impossible to destroy (for example, anthrax). In this situation military commanders face real dilemmas.

'Operation Desert Fox' launched in December 1998 against Iraq saw some of these dilemmas played out. The ostensive reason for launching the air strikes was Iraqi non-compliance with the UNSCOM inspection teams, particularly over the issue of their BW programme. Air strikes by both manned bombers and cruise missiles were launched with the stated aim of degrading the regime's CBW capabilities. However, it subsequently emerged that the British and American forces had not targeted known storage sites for fear of releasing CBW into the atmosphere. It was claimed during the operation that the allies would target the factories which were suspected of manufacturing CBW, but even this carried the dangers of dissemination. As a consequence, the majority of the targeting related to delivery systems, hitting missile and unmanned aerial vehicle (UAV) factories and storage sites. It was not clear at the close of the bombing that any of the Iraqi regime's CBW had been degraded, although the manufacture of means of delivery had certainly been retarded. Ironically, rather than making Britain and the USA look powerful, the problems of the strategic bombing campaign made them look weak and ineffective.

Systematic destruction

As noted above, the UNSCOM operations in Iraq are the best recent example of a systematic destruction operation. It is being carried out under particularly difficult circumstances, owing to the passive and active non-cooperation of the Iraqi authorities. Disputes with Iraq have thus far prevented the completion of the inspection and destruction process.

Disarmament

Other examples of systematic destruction include the United States and Britain's unilateral disarmament of their BW programmes in the wake of the establishment of the BTWC. However, the reversibility of disarmament has to be borne in mind, with both retaining defensive programmes and having access to civil manufacturing facilities which could be readily turned to creating BW. It is also worth noting that both programmes have had lasting environmental consequences.

Conclusions: implications for international relations

The use of BW is a terrifying prospect. Consequently, states have been working to forestall the possibility of this occurring. However, the already difficult hurdles to be overcome, involving issues such as verification and legitimate defensive developments are being augmented by new dilemmas brought on by the scientific advances being made in the fields of molecular biology, genetic research, etc. Consequently, more states are realising that denial strategies are not enough, that they need to be augmented by efforts to deter use, detect production and use, defend populations, animals and crops and destroy potential BW capabilities.

The BW threat is also a significant challenge to the way in which international relations as a discipline has traditionally regarded threats to security. Until recently, most of the IR literature looked at security issues in terms of the state alone. This is an inadequate way to view the BW problem for five reasons. First, the BW problem is a transnational issue whose effects would not be physically or politically limited by state borders. Secondly, the BW problem requires transdisciplinary solutions (for example the co-operation of microbiologists, gene therapists, etc.) which defies the standard 'closing off' of issues as purely military concerns. Thirdly, the biotechnology revolution is subverting existing regulatory regimes, not just in the realm of security, but, as this volume has shown, in many different areas including the environment, food, ethics, etc. This means that regimes are running to try to keep up with the pace of developments – a challenge to regime theory. Fourthly, tackling the problem of BW requires the co-operation of many different actors in addition to the state; international organisations, private firms, doctors, scientists, etc. This is a particular challenge to the state-centric approach to security. Finally, future trends in the BW problem are dependent on trends in other disciplinary areas, for example agriculture and mono-cropping. For all these reasons, international relations scholars will be forced to think 'outside the box' when they consider the biological weapons problem.

14

Conclusion: the implications for international relations

Alan Russell and John Vogler

The impact of biotechnology on international relations is inherently complex. While the underlying science is built upon focused new knowledge and specialist techniques the ramifications of the ability to manipulate DNA are widespread, complex and profound. DNA, it must not be forgotten, is the fundamental basis of *all* life. Biotechnology is a generic technology. The same basic techniques underpin activity along many pathways. In turn these pathways thread their way through the concerns of International Relations. The complexity of biotechnology confronts the complexity of International Relations. In academic terms it is readily evident from the differing ways in which the problem is 'framed' by students of security or IPE. Security analysts regard biotechnology as providing the potential for novel weapons in a largely state-centric conception of the world. They are concerned about the usability of such weapons and the ways in which state authorities may or may not be able to defend against them. IPE concerns are altogether different. They will include the implications of a rapidly growing new biotechnology industry for the structure of the international economy and questions of national competitiveness. At a more profound level there will be an awareness of the pervasive effects of technological change and the need to understand the innovation process and to reconceptualise classical models of the international economy to include a knowledge dimension.

The implications of such varying perspectives can be illustrated through the example of the way in which the new biotechnology will bear upon the power of countries of the South. For students of IPE the development of the global biotechnology industry will accentuate the existing weaknesses of less-developed economies in the face of the dominance of Northern firms. On the other hand, security analysts will point out the relatively low cost availability of possible weapons of mass destruction, which may be acquired with a modest technological capability. On some issues the two perspectives converge. Both contain a view of the international system which seeks order from or within an anarchic states system through the development of international co-operation and regulatory regimes. Hence the attempts to ensure the rights to intellectual property on an international basis, to standardise bio-safety regulations and to erect a biological weapons control regime. Both are concerned about the continuing role of state authority in the face of the trend to globalisation. Ultimately the discourses contained in these two perspectives may neglect an underlying cultural and value dimension which links people to the global system (Dyer, chapter 3, this volume; Bretherton and Stevenson, chapter 4, this volume).

The preceding discussion suggests we might usefully organise our summary of the implications of biotechnology for international relations along the following lines:

- security;
- the structure of the international political economy;
- globalisation effects and the erosion of state autonomy;
- culture and resistance;
- regimes and global governance;
- international relations theory.

Biotechnology and security

As a field of study international relations has directed an enormous, perhaps inordinate, proportion of its attention to issues arising from the development of weapons of mass destruction. This was particularly so within the US community of scholars whose work mirrored a national policy concern with the Cold War arms race. The links between biotechnology and weaponry are growing ever closer. While many of the classical agents identified by Dando are not subjected to any molecular manipulation, when weaponised they become part of a traditional human endeavour to utilise biological materials. It is, therefore, important to understand the consequences of these endeavours as a precursor to discussing ways in which the new biotechnology might influence weapon development. The Biological and Toxin Weapons Convention and its review process must account for old and new possibilities and the problem of 'dual-use' technology, including genetic manipulation techniques, as well as the growing knowledge base (Littlewood, chapter 12, this volume). Equally, concerns about state security may encourage reliance on deterrence or denial strategies, for example, whatever the precise technological basis of the biological weapons threat that might be faced (Spear, chapter 13, this volume). There are various ways to develop biological weapons and, therefore, associated classifications. Some may replicate and infect people, animals or plants – and may be agents like anthrax or some genetically engineered organism. Yet others may be toxins, which do not replicate but have biological origins, where perhaps new biotechnology techniques can help in their production or extraction. The real issue is *uncertainty*. Uncertainty about who is developing biological weapons at all, and uncertainty as to their nature – partly as a consequence of the secrecy that surrounds these matters. State authorities may well have an interest in reducing the likelihood of an enemy developing vaccines or other counter measures.

Biological weapons are relatively cheap compared with nuclear armaments and technologically refined weapons; using genetic engineering techniques are readily available for those with the requisite technological capacity. Thus, states may seek to develop biological weapons as a more achievable alternative to nuclear weapons. Such weapons may then enter the arsenal of destruction potentially at the state's disposal. The study of international relations must take such developments on board as part of a long-standing concern with the means of war. A novel problem, however, is the potential acquisition of biological weapons by non-state actors. Such weapons in the hands of terrorist groups represent a new order of threat (Spear, chapter 13, this volume) that state-led deliberations on a biological weapons regime must attempt to address. The difficulty, as Littlewood reminds us, is that the interdisciplinary dual-purpose knowledge involved is so widespread, that any 'simple' solution, satisfactory to all, is unlikely. Moreover there is the additional problem of attempting to construct a control regime when the progress of the technology is likely to outpace the development of regulation. Comprehensive involvement of governments and end-users of biotechnology may offer some hope that a wave of pre-emptive strikes against those suspected of developing biological weapons can be avoided.

The structure of the international political economy

A number of large-scale studies which investigated the growth of biotechnology, especially in the 1980s, drew attention to the competitive strengths of the United States, Japan and the collective states of Western Europe (US Congress, OTA, 1991; Yuan, 1987; UN Centre on Transnational Corporations, 1988). In effect, the industrialised world has taken the lead in developing biotechnology, although its long-term impact will be global. As biotechnology has its main impact as an enabling technology, often assisting older industries (including agriculture), the competitive status of the industrialised countries is derived both from traditional strengths and from their ability to develop and utilise the new biotechnology. Consequently, there is much divergence in the characteristics of national biotechnology strategies and corporate interests. This divergence extends to 'technology transfer' from academic research institutions to industrial users.

The United States has the widest range of interests and the strongest overall research base. Government funding also outstrips that of its rivals and this is likely to help maintain its research lead. However, the US government is not pursuing any clear industrial policy for biotechnology, although it does have a tradition of mission-oriented policy usually centred on 'big science' projects, and often with a security dimension. Biotechnology does not fit the traditional big science image as it is so diffuse in its applications – despite the Reagan Administration's classification of biotechnology as a strategic technology.

The perceived lead and success of American biotechnology has drawn a competitive response from other industrial countries, notably Japan. The growth of NBFs in the US by the 1980s and the equity investment they were attracting, combined with the apparently optimistic US forecasts for biotechnology, helped engender Japanese government interest (Saxonhouse, 1986: 98–9). Unlike the US approach, the Japanese approach to biotechnology has clearly been more targeted (US Congress, OTA, 1991: 153–8; Committee on Japan, 1992). Here it is the largest companies – supported by universities, research centres and government bodies – that dominate biotechnology development (UN Centre on Transnational Corporations, 1988: 55; Kinoshita, 1993).

But compared with the United States, Japan's publicly funded basic research is weak. Instead, its strength is in industrial research and development. In general, industrial researchers active in natural science are twice as numerous as government and university researchers. By the early 1980s the Japanese government was funding around 20 per cent of biotechnology basic research compared with the US government, which funded around 50 per cent. However, such comparisons under conditions of globalisation are hardly straightforward. The degree of transnational collaboration between firms often confounds the intentions of national policy makers.

Assessments of European competitiveness in biotechnology must embrace individual European countries as well as the integration of the region and its position in the world economy (Pownall, chapter 7, this volume). European biotechnology has many strengths at both the applications level and at the level of basic research. However there are some important contrasting characteristics of individual countries. Much French biotechnology has been state initiated, following a 1982 Mobilisation Programme striving for 10 per cent of the world's biologically based production by 1990 (US Congress, OTA, 1991: 159; Yuan, 1987: Appendix A, 1–31). Germany, in general terms, has a strong industrial base and many of its large chemical and pharmaceutical companies have, since the early 1980s, utilised the new biotechnology techniques. Support came from the Federal government. However, a particular problem for Germany has been a relatively high level of distrust regarding biotechnology amongst the general population, with much attention coming from the Green Party (Toro, 1992a; 1992b). In contrast, industrialists have mounted a challenge to the bureaucracy surrounding the licensing of gene experimentation following the Gene Law passed in 1990. Consequently, many German firms have carried out their biotechnology research outside Germany.

The United Kingdom has followed an academic-led biotechnology development path. The UK approach is closest to that of the United States, in that the government has been disinclined to develop a clear focus for biotechnology while there is a substantial underlying research base. The UK does, however, have some large multinational companies that are active in biotechnology research and development, including ICI and Glaxo Wellcome, and there have been important start-ups (Lynn, 1993). Foreign firms value the UK's basic research in biotechnology, which, for example, led Monsanto to enter into a £20 million deal with Oxford University (US Congress, OTA, 1991: 160). The big European players are completed by Switzerland with its large chemical and pharmaceutical companies, such as Sandoz, Ciba-Geigy and Hoffman-La Roche.

There are deep and complex processes at work – only some of which reinforce state power. They can be analysed via Susan Strange's concept of a 'knowledge structure' where structural power has intentional and non-intentional characteristics (Strange, 1994; Guzzini, 1993; Russell, chapter 6, this volume). The US lead is not a consequence of US government policy, no matter how supportive. The rest of the world has found itself responding in various ways to the structural influence of the US economy, which lies at the centre of many transnational links and networks generating a cumulative and non-intentional impact. In this sense it is worth noting the way that social controversy in the USA, over the safety of genetic engineering, was left behind in the 1980s, smoothing the path for subsequent R&D and the widespread exploitation of the technology (Wright, 1994). In the late 1970s the two countries which had led the efforts to develop safety procedures were the USA and the UK. For the USA it was a highly charged affair under attentive public and media scrutiny.

However, in the US the discourse changed. Against a backdrop of a general trend towards deregulation, 'corporate representatives and scientists alike framed the genetic engineering "problem" in terms of a "race" in which the ... controls represented a major "handicap". Scientific interests in genetic engineering had been profoundly restructured by this point' (Wright, 1994: 451). Hard won controls were gradually relaxed setting the scene for a massive expansion in US biotechnology. Even fear of biological weapons development did not slow things down, as the scientific community and the immediate regulators tended to side-step the issue, neutralising any linkage. The importance of the US leaving behind the safety issue is given added poignancy with the growing European concern over environmental release of GMOs, and fears over GM foods.

There is a real sense of competitiveness involved as governments strive to protect the position of their national science and technology bases. This, of course, brings them into interaction with firms, while in turn firms network into the wider global system (Pownall, chapter 7, this volume). Inevitably, the structure of the international political economy, and biotechnology within that, is affected by factors other than government policies, as the debates on safety in the 1970s and more recently in Europe in the late 1990s testify.

Globalisation effects, the erosion of state autonomy and non-state actors

The traditional state-centric frameworks of International Relations struggle ultimately to accommodate the new biotechnology, both in policy terms – witness the regime difficulties over both the environment and biological weapons – and in terms of any state-centric ontology as a basis for IR theory. A full understanding of the impact of biotechnology is only possible when we think in terms of the processes of globalisation.

Technology transfer, the activities of firms in biotechnology, consumer concerns, ethical issues and firm-government interactions all include transnational features. The process of globalisation encourages such linkages and the development of biotechnology – as a high intensity transnational technology – can be argued to add to globalising trends. The associated erosion of the state can only be evaluated through analysis of all the drivers of globalisation, including technology, and from careful consideration of the influence and interactions of non-state actors.

US firms, for example, have found themselves at the centre of increasingly complex links with Japanese firms (such as Kirin-Amgen, Kirin-Calgene, Showa-Toyo Diagnostics-Monotech). Japanese efforts have involved tapping into the scientific strength and expertise within the USA. This is symptomatic of the globalisation of business activity and in this case reflects both collective US structural influence, in the transnational network sense, and relative influence as a nation. Moreover, as argued above, the structural influence is often non-intentional rather than reflecting a clear US policy. Nevertheless it is likely that the US government will strive to maintain a general structural influence, through its leadership in basic research and its ability to influence international regimes, while suffering some set-backs in relative competitiveness and influence. However, the identification of 'national' interests in circumstances of globalisation is not a straightforward task (Reich, 1991). It is further complicated by second guessing where international challenges to US dominance might arise in future. For example, Japan's challenge has been overstated. Weaknesses in basic science, the failure by and large of Japanese firms to bring biotechnology products to market, its inability to develop and refine the genetic tools that underpin biotechnology and its dependence on biotechnology imports, leave it in a clear second place compared with the vibrancy of US research and the number of NBFs maturing there (*The Economist*, 1995).

As well as intra-European links, there are also links between European firms and those of the US and Japan (Cantley and de Nettancourt, 1992;

Prevezer and Toker, 1996). It would thus seem that the biotechnology industries of both Japan and Europe depend to a considerable extent on transnational linkages with US firms, which are in turn plugged into the thriving US science base. To counteract this, the EU has attempted to influence the structural framework within which intra-European biotechnology progresses (Pownall, chapter 7, this volume). In support of the individual countries of Europe, there have been efforts to enhance key biotechnology trajectories, notably in relation to pharmaceuticals and agriculture, and thus shape the character of the developing technological paradigm itself.

Alliances of firms bargain with individual or collective governments (such as the lobbying of biotechnology firms to achieve international intellectual property protection and unrestricted access to biological diversity – Williams, chapter 5, this volume). Firms of course also bargain with each other, often across borders (as Japanese firms have entered liaisons with US biotechnology firms). More interestingly, mixed alliances of firms and governments increasingly face other similar mixed alliances – something evident in the background context of the Rio Summit, the TRIPS negotiations of the Uruguay Round, and in the complex lobbying process of the European Union. In all three cases biotechnology has been prominent. The ability to shape structures, in the terminology of Strange (1994), can be sourced to such mixed groupings whether it is deployed with intent or unintentionally. Although when intent is evident we may be witnessing what Stopford and Strange describe as the 'new diplomacy' of relations between complex groups of firms and states (Stopford and Strange, 1991; Pownall, chapter 7, this volume) – relations that may be influencing the shape of new technological paradigms and systems.

There are far fewer start-up biotechnology firms in Europe compared with the United States. In large measure it is the big firms setting the pace in Europe. Large pharmaceutical companies, which in Europe also tend to be the dominant chemical firms, are particularly important. Nevertheless, there has been a degree of caution in European companies adopting the new techniques of biotechnology (Pownall, chapter 7, this volume).

Many NBFs simply have not had the resources to bring their products to the market. In the pharmaceuticals sector, typically, the cost of meeting regulatory requirements proved excessive, or the risk of failure unacceptably high. The large firm could also have a substantial international marketing operations, impossible for the small firm to develop. Or it may simply be the case that the NBFs have found funding harder to acquire as the optimism of the early days appears to have been misplaced. Whatever the reason, many alliances have emerged or, alternatively, the NBF has been bought out by a larger firm (Ng, Pearson and Ball, 1992; *The Economist*, 1991; Sharp, 1991; *Business Week*, 1992; Van Tulder and Junne, 1988).

These linkages are, therefore, often transnational. They also reflect the globalisation of the world economy with European–US and Japanese–US collaborations (Kinoshita, 1993; *Chemistry and Industry*, 1992). The thrust of the Single European Act was to reduce barriers to competition *within* Europe, making transatlantic collaboration attractive, in contrast to the objectives of the Technology Framework programmes (Pownall, chapter 7, this volume). In agriculture a series of large firms, including ICI have purchased established, but smaller, seed companies (Williams, chapter 5, this volume). Here the intention is to acquire expertise as well as outlets for genetically modified seeds in world markets. As Hobbelink has observed:

> Yet even before anyone thought of biotechnology as the driving force behind the restructuring of agriculture, TNCs started massively buying up independent seed companies. With or without biotechnology, the seed is the ultimate vehicle of genetic improvements in agriculture, and the TNCs realised the importance of that. Once a highly diverse and often family-based sector with no TNC involvement, virtually all top seed houses now have their main interest in either chemicals, pesticides or pharmaceuticals. (Hobbelink, 1991: 45)

Whatever the significance of the NBFs, and for the last decade there has been a constant twelve hundred of them in the US (NgPearson and Ball, 1992: 352; Gaull, 1997), in the long term, developments in biotechnology at a global level will increasingly be dominated by large and generally multinational firms. One consequence of this is that these firms can command considerable individual or collective influence on the policy process of states. Another consequence is that their activities are of a scale that will shape North–South relationships in biotechnology concerns, including biodiversity, agriculture, medical treatment and pharmaceuticals.

Biotechnology, culture and resistance

Globalisation has broader associations than the activities of firms and other transnational actors. Ideas, beliefs, culture and knowledge are all involved. Much of what has come to be identified with globalisation is less tangible (Dyer, chapter 3, this volume). For biotechnology, cultural differences have brought varied take-up rates in the development of biotechnology products and consumer response. Youngs suggests that a tendency in the literature has been to view the relationship between technology and culture in hierarchical terms. Culture becomes the servant or object of technology 'because technology is rational, and culture irrational' (Youngs, 1997: 34). Culture is characterised as nationally based – a defining characteristic of the nation itself – while technology is transnational and global. The dominance of technology becomes associated with an ideology of global progress and coalesces as a basis of global culture. Fukuyama encapsulates such a view:

> The enormously productive and dynamic economic world created by advancing technology and the rational organisation of labour has a tremendous homogenising power. It is capable of linking different societies around the

world to one another physically through the creation of global markets, and of creating parallel economic aspirations and practices in a host of diverse societies. (Fukuyama, 1992: 108)

Such views recognise – and often advocate – the dominance of a capitalist ideology linked culturally to consumerism, and supported by an emphasis on the individual in a liberalised context. Biotechnology plays its part in this. Yet, such views are eminently challengeable and Bretherton and Stevenson (chapter 4, this volume) are among those who do not accept the separation of culture and technology. The Actor-Network conceptualisation, too, can help bridge this apparent divide by encompassing heterogeneous relationships between humans and non-human artefacts – the producers and products of technologies (Russell, chapter 6, this volume).

Aside from possible structural effects in the global political economy there are broad issues of the social acceptance of certain biotechnology practices, such as cloning (perhaps to include human cloning in certain circumstances), genetic screening of employees, gene therapy, genetically modified foods and the reconciliation of concerns over 'interfering with nature'. The issue of intellectual property protection (Williams, chapter 5, this volume) prises open some of these concerns. If large countries, perhaps with a collective societal view, can push through a regime for the international recognition of the patenting of life, then this may be at the expense of offending the sensibilities of other societies. Commercial interests do influence the way in which technologies in turn are accepted by societies. Developing global consumerism, within large sectors of the world's population, may drive companies to deliver the products of biotechnology.

But there is also resistance (Falkner, chapter 9, this volume). A growing network of farmers in India has started to take on the seed companies – themselves networked at a different level; for example in relation to campaigning for intellectual property protection. Vandana Shiva has a leadership role in the growing mobilisation of resistance in India, and has observed: 'What the Europeans have rejected due to environmental and health hazards is now being dumped on Indians who traditionally do not consume soya' (quoted in IPS, 12 August,1998). The context of Shiva's words was a rally which aimed to prevent the landing of one million tonnes of US soyabeans, suspected to be genetically engineered. Resistance to genetically engineered seeds is also evident from India's agricultural scientists' search for the 'Terminator gene' developed by US biotechnologists. This, they argue, threatens the livelihood of 400 million farmers and hence food security in India. 'We will not allow the Terminator to enter this country', Dr R. S. Paroda, director-general of the prestigious Indian Council of Agricultural Research (ICAR) told IPS (IPS, 15 July, 1998).

The growing resistance in the South is also evident in Latin America. An IPS report on the first Latin American and Caribbean Indigenous Seminar observes: 'Latin American farmers now buy seeds made in labs in the North from genetic material they donated in the 1970s, just as South American nations imported British goods manufactured from their wool and leather in the 19th century' (IPS, 27 August 1998). Moreover, the same report suggests that 'more than half the known plant species in Brazil, one of the countries with the richest biodiversity in the world, have already been patented by large transnationals'. One transnational, it is claimed, went even further and took blood samples from Mexico's Yaqui Indians to extract and synthesise an antigen produced naturally by their bodies. As Bretherton and Stevenson (chapter 4) argue, commodification of human (and other) genetic material has assumed an explicitly North–South dimension.

Networks of resistance developing in India and elsewhere are becoming reinforced by rising resistance in developed countries, often articulated with the help of transnational pressure groups. Most notably this is happening in Europe with a mix of a consumer backlash and direct action – the latter, for example, involving the destruction of GM crops planted in field trials. Britain has been a centre for the debate, leaving the UK government struggling to articulate a policy on the issues raised. However, the most significant reaction might ultimately be that in the US, where GM crops and food to date have had a relatively smooth ride. Thus, despite estimates that genetically engineered corn acreage planted in 2000 would be 20 per cent higher than in the previous year, Gary Goldberg of the American Corn Growers Association told the Washington Post that GM corn acreage in the US was likely to be *down* by as much as 25 per cent in 2000. Goldberg suggested that farmers were seriously concerned over potential loss of overseas markets for such crops, especially in Europe (UPI, 12 September 1999). Despite the earlier smooth ride 'worries about potential health and environmental effects overseas has prompted some major food processors to declare they won't buy such crops. The boycott leaves farmers who grow engineered crops with fewer potential customers' (UPI, 12 September 1999).

The serious reverses suffered by large companies, such as Monsanto, as a consequence of the increasingly global resistance to GM products may soon be exacerbated by a further, peculiarly American, response. A multi-billion dollar antitrust lawsuit is being prepared for launch in up to thirty countries. Targeting companies like Monsanto, DuPont and Novartis, it will claim that they are exploiting bioengineering techniques to gain a stranglehold on agricultural markets. Twenty US law firms have signed up on a no-win no-fee basis. The action is being brought jointly by the Foundation on Economic Trends, run by Washington-based biotech activist Jeremy Rifkin, and the US-based National Family Farm Coalition, together with individual farmers across Latin America, Asia, Europe and North America (*Financial Times*, 13 September 1999). The companies are sure to fight this lawsuit fiercely as the future global industry may be at stake. Against such developments it is interesting to note that Monsanto has become involved in a merger that will see the 'Monsanto' name applying only to its agricultural concerns.

In a neo-Gramscian conceptualisation, resistance to the transnational seed industry may be symptomatic of a growing movement of consumers and small producers in opposition to the transnational hegemony of corporate interests that dominates the global economy. A true counter-hegemony may be a long way off but few issues in modern times have created such a response from 'consumers', both in the North and in the South. This facet of globalisation – the rapid spread of a new, pervasive and generic technology – has profound implications. It impacts upon many aspects of life and life processes at the human level. Most fundamentally, commodification of the processes of birth and death is experienced by many as deeply troubling. Fears concerning the consumption of unnatural and unsafe food provide a further significant factor influencing opinion; as does uncertainty over the consequences of releasing GMOs into the environment. While these concerns are experienced and articulated differently in different cultural contexts, they link sites of resistance in the North and the South, presenting an unprecedented challenge to the premises of universalised scientific discourse. Unprecedented, too, is the challenge to governance at the international level.

Biotechnology and governance

In turning our attention to the question of governance, we find a key contradiction between, on the one hand, the evident requirement to construct international regulatory arrangements for biotechnology and, on the other, the conflicting demands of transnational capital and increasingly mobilised popular opinion (Falkner, chapter 9, this volume).

The requirement for international governance stems from developing global markets in genetically engineered products and, in particular, the potential for deleterious health and environmental consequences of unregulated transfers of GMOs across frontiers. At present, despite serious gaps between the existing regimes for trade, environment, health and food safety, a number of factors impedes construction of a broadly based coalition capable of relating and consolidating these islands of regulation. So formidable are the challenges facing regulation of biotechnology that they apply even in the relatively mundane area of aligning existing national safety regulations.

The incompatibility of the interests involved constitutes a central barrier to construction of an adequate international governance system for biotechnology at the formal level. The corporate sector demands regulation to protect investments in intellectual property in an environment which remains open to trade in GMOs. Southern governments have an interest in rules that reinforce their control over what are regarded as sovereign genetic assets. Environmental NGOs and consumer groups, as well as small-scale agricultural producers in both the North and South, are demanding a moratorium on production and transportation of GMOs. Failure to make progress in constructing formal regulatory mechanisms, however, should not obscure the emergence of informal norms and principles associated with conflicts over biotechnology. There is substantial evidence that the behaviour of many of the principal biotechnology firms is now being affected by the informal constraints imposed by public resistance, and the consequent normative shift against the acceptability of GMOs.

The regime literature tends to concentrate upon the negotiated construction of formal agreements between governments. A central finding of many of the chapters of this book is that development of biotechnology is led by the private sector. This is true even in relation to arms control, where the principal difficulty appears to arise, not in constraining governmental behaviour, but in preventing proliferation to non-state groups. In comparison with existing environmental and public health regimes, there are particular regulatory difficulties when governments confront biotechnology. There are, as yet, no damaging emissions to abate. For opponents, the demand is not for production of safer alternative products, rather there is a call for a return to 'natural' production. For advocates, this does not represent a sustainable option in terms of future global food supplies. Moreover the potential for beneficial use of biotechnology in environmental protection and areas of medicine would be lost. As Manning has demonstrated, there have been many important and beneficial applications of biotechnology.

A further impediment to regime construction in relation to biotechnology is the uncertainty, even ignorance, which permeates the whole debate. This contributes to the loss of authority of official science in areas such as food and environmental safety. Thus, biotechnology lacks the relatively consensual scientific underpinning that has been regarded as fundamental to the construction of environmental regimes. This also makes it difficult to apply the precautionary principle without yielding entirely to demands for a moratorium – hence the considerable difficulties in accommodating safety measures for biotechnological products within the existing disciplines of the GATT/WTO regime.

The final, and most significant, challenge to international governance posed by biotechnology flows from its evident ability to link the local and the global. The impact of biotechnology upon processes and products fundamental to human life are immediate – and, in consequence, highly susceptible to politicisation.

Biotechnology and international relations theory

This book began with an expression of discontent at the way in which technological change had been regarded as an 'exogenous' variable by most students of IR. The contributors to this study consciously set out to place biotechnology at the centre of their analyses and to consider some of the implications of what is widely regarded as *the* twenty-first-century technology (Arber and Brauchbar, 1998). There was no attempt to stake out a new theoretical position. Rather, the essays in this volume largely represent an attempt to consider the new technology in terms of existing theorisation in IR. They utilised approaches such as regime analysis, structural perspectives on IPE and security conceptualisations. Nevertheless, many of the contributions illustrated the benefits of adopting an interdisciplinary approach, in a dialogue between natural and social science, in business-oriented approaches to the internalisation of technology and in the application of Actor-Network Theory.

The international implications of biotechnology are not simply grist to the mill for established theoretical approaches. The chapters in this volume indicate that the relevance of more traditional state-centric frameworks is limited. This is hardly novel. Many scholars in IR have become discontented with the primacy of state centricity, or, indeed, structural approaches emphasising states. Such discontent has been evident from the mid 1970s and has been associated with efforts to broaden the issue agenda of IR (Keohane and Nye, 1977); and from perspectives which question the prevalent processes of global interaction and the contextualisation of structures (Held *et al.*, 1999; Strange, 1996; Halliday, 1994).

The essentially transnational nature of biotechnology has been a *leitmotiv* of this volume. We have identified problems raised in dealing with the potential spread of biological weapons across national borders and into the hands of non-state actors. In terms of the international political economy, it is clear that the structure of the biotechnology industry is highly transnational. In both of these contexts it is important to recognise that states are not the only actors of importance, and that power cannot be seen in simple state-centric terms nor in hierarchical structural terms without an account of agency. This leaves a need to extend further the range of actors and actions in our considerations. Actor-Network Theory, for example, points a way to building up to the macro from networks of agency with origins in localised micro activity Moreover, this extension offers scope to acknowledge challenges to conceptualisations of hegemony, whether viewed in hierarchical state centric terms (Gilpin, 1987) or in transnational neo-Gramscian terms (Cox, 1987).

While much effort in IR theory has been put into the analysis of regimes, this has tended to occur within very specific dimensions, such as trade or money. The study of the new biotechnology and its various transnational ramifications alerts us to the interface between established issue areas and regimes – trade, environment and security. It also, most significantly, points out their incongruity and incompatibilities. This is most clearly illustrated in the policy dilemmas associated with the tensions between the application of WTO and CBD inspired rules on the handling of living modified organisms. An extension of regime theorising in this direction would clearly be desirable.

Many of the contributions in this volume reinforce a trend towards explicitly incorporating technology as an independent explanatory variable within mainstream theorisation (Skolnikoff, 1993; Rosenau, 1990). While this was and remains an important theoretical task within IPE, the unanswered questions thrown up by the consideration of biotechnology may excite as much, if not more, interest. For example, how can we establish the patterns of vulnerability associated with the biotechnological revolution? Who will benefit, who will be endangered and how might security be re-conceptualised? Is it necessarily the case that LDC's will always be most vulnerable especially as they lack well-developed public health systems. Related and very immediate questions are posed by some of the reactions to GMOs and biotechnology in general recounted in this volume. There appears to be a special fear and dread associated with aspects of biotechnology that were at other times and in other circumstances regarded as mundane and even benign. To account for such politically significant reactions we need to delve into cultural and psychological factors far removed from orthodox IR.

IR has often been described as an eclectic discipline, borrowing extensively from other subject areas – be it economics, sociology, law, political economy or philosophy. Thus behavioural theories of the firm came to influence foreign policy analysis; micro-economics provides a basis for Waltz's structural realism (1979); sociology informed the World Society perspective (Burton, 1972). More recently, there has been a tendency towards interdisciplinary discourse affecting many areas of study, thus raising questions concerning the boundaries of IR (MacMillan and Linklater, 1995). If IR is to move towards embracing technology in an explicit fashion it should consider these other contributions and their theoretical orientations. Just as the antipathy and resistance generated by biotechnological applications has challenged hegomonic discourses of scientific progress and economic globalisation, so it has opened up new spaces for interdisciplinary discussion – between and beyond the social sciences.

Bibliography

Adam, M. (1997), 'Entrepreneurialism and the European Commission', *Nature*, Autumn, 9–12.

Adams, L. G. and J. W. Templeton (1998), 'Genetic resistance to bacterial diseases of animals', *Epizooties*, 17, 1: 200–19.

Aldridge, S. (1996), *The Thread of Life: The Story of Genes and Genetic Engineering*, Cambridge, Cambridge University Press.

Alibeck, K. and S. Handelman (1999), *Biohazard*, London, Hutchinson.

Alic, J. (1993), 'Technical knowledge and technology diffusion', *Technology Analysis and Strategic Management*, 5, 4: 369–83.

American Kennel Club (1997), *The Complete Dog Book*, 19th edition revised, Howell Book House.

Amin, A. and R. Palan (1996), 'Editorial: the need to historicize IPE', *Review of International Political Economy*, 3, 2: 209–15.

Andersen, A. L., B. W. Henry and E. C. Tullis (1947), 'Factors affecting infectivity, spread, and resistance of *Piricularia Oryzae* Cav', *Phytopathology*, 37: 95.

Anderson, A. (1988), 'Oncomouse released', *Nature*, 336, 6197: 300.

Anderson, J. (1996), 'Feeding a hungrier world', *Phytopathology News*, 30, 6: 90–1, available from Promed@usa.healthnet.org

Andrews, E. L. (1998), 'Europe's once stodgy investors take a fling on high-tech stocks', *The San Diego Union-Tribune*, 28 June, I-5.

Anonymous (1996), 'Criminalizing BW', Editorial, *Chemical Weapons Convention Bulletin*, Issue Number 31, March 1996.

APEC (1998), *1998 APEC Economic Outlook: Economic Trends and Prospects in the APEC Region*, Economic Committee Asia-Pacific Economic Co-operation, November.

Arber, W. and M. Brauchbar (1998), 'Biotechnology for the 21st Century', in OECD, *21st Century Technologies*, Paris, OECD.

Archibugi, D. and J. Michie (eds) (1997), *Technology, Globalisation and Economic Performance*, Cambridge, Cambridge University Press.

Badaracco, J. (1991), *The Knowledge Link*, Boston, MA, Harvard Business School Press.

Bailey, K. (1995), 'Responding to the threat of biological weapons', *Security Dialogue*, 26, 4: 383–97.

Bains, W. (1987), *Genetic Engineering for Almost Everybody*, Aylesbury, Penguin.

Bains, W. (1996), 'Company strategies for using bioinformatics', *Trends in Biotechnology*, 14, 8: 312–17.

Balmer, B. (1997), 'The drift in biological weapons policy in the UK 1945–1965', *Journal of Strategic Studies*, 20, 4: 115–45.

Balmer, B. and M. Sharp (1993), 'The battle for biotechnology: scientific and technological paradigms and the management of biotechnology in Britain in the 1980s', *Research Policy*, 22: 463–78.

Barbosa, D. (1995), 'Letter from the Gamma World', *International Journal of Technology Management*, 10, 2/3: 230–6.

Barnum, A. (1993), 'FBI probes spying on biotech firms', *The San Diego Union-Tribune*, 10 July, C-1.

Bartfai, T., S. J. Lundin and B. Rybeck (1993), 'Benefits and threats of developments in biotechnology and genetic engineering', *SIPRI Yearbook 1993: World Armaments and Disarmament*, Oxford, Oxford University Press and the Stockholm International Peace Research Institute.

Bartlett, J. T. (1996), 'The arms control challenge: science and technology dimension', Paper presented at a NATO Advanced Research Workshop on 'The Technology of Biological Arms Control and Disarmament', Budapest, 28–30 March.

Barton, J. H. (1995a), 'Adapting the intellectual property system to new technologies', *International Journal of Technology Management*, 10, 2/3: 151–72.

Barton, J. H. (1995b), 'Patent scope in biotechnology', *IIC*, 26, 5: 605–18.

BBC, Today (1999), Radio 4 *Today* programme, item on the Russian BW programme, 28 April.

Beckerman, W. (1992), 'Global warming and international action: an economic perspective', in Hurrell and Kingsbury (1992), pp. 253–89.

Beier, F. K. and G. Schricker (eds) (1989), *GATT or WIPO? New Ways in the International Protection of Intellectual Property*, New York, ICC Studies.

Beitz, C. R. (1983), 'Cosmopolitan ideals and national sentiment', *Journal of Philosophy*, 80, 10, October.

Beresford, D. (1998), 'S Africa "Sought germs to attack blacks only"', *Guardian*, 12 June, 15.

Beresford, D. (1999), 'South Africans were working on blacks-only germ', *Guardian*, 22 January.

Berliner U. (1999), 'Venture capital investment here strong despite dip', *San Diego Union-Tribune*, 16 February, C-1.

BHRC – BIOTEC Canada Human Resource Council directory listing of current University, College, Schools and Skills programme (1998), sourced, April 1999, from http://www.biotech.ca/main.html

Bifani, P. (1989), *Intellectual Property Rights and International Trade*, New York, UNCTAD, UNCTAD/ITP.

BioAsia Monitor (1998), 'General profile of biotechnology activities in Asia', sourced at www.biocompass.com/ April 1999.

Biology Department (1950), 'Feathers as Carriers of Biological Warfare Agents: I. Cereal Rust Spore', Special Report No. 138, Chemical Corps SO and C Division, Camp Detrick, Frederick, Maryland.

Biotechnology Directorate (1992), 'Biotechnology directorate: forward plan', Swindon, Science and Engineering Research Council.

Birnie, B. (1992), 'International environmental law: its adequacy for present and future needs', in Hurrell and Kingsbury (1992), pp. 51–84.

Birnie, P. W. and A. E. Boyle (1992), *International Law and the Environment*, Oxford, Clarendon Press.

Blackhurst, R. and K. Anderson (eds) (1992), *The Greening of World Trade Issues*, Hemel Hempstead, Harvester Wheatsheaf.

Boily, M. C., B. R. Masse, K. Desai, M. Alary and R. M. Anderson (1999), 'Some important issues in the planning of phase III HIV vaccine efficacy trials', *Vaccine*, 17: 989–1004.

Bordo, S. (1993), *Unbearable Weight: Feminism, Western Culture and the Body*, Berkeley, University of California Press.

Bourdeau, P., P. M. Fasella and A. Teller (eds) (1990), *Environmental Ethics: Man's Relationship with Nature; Interactions with Science*, Luxembourg, Commission of the European Communities.

Brack, D. (ed.) (1998), *Trade and Environment: Conflict or Compatibility?* London, Earthscan.

Braga, C. P. (1996), 'Trade-related intellectual property issues: the Uruguay Round and its economic implications', in Martin and Winters (eds) (1996), pp. 341–79.

Breman, J. G. and D. A. Henderson (1998), 'Poxirus dilemmas – monkey pox, small pox, and biological terrorist', *The New England Journal of Medicine*, 339, 8: 556–9.

Brenner, C. (1997), 'Biotechnology policy for developing country agriculture', Policy Brief No. 14, Paris, OECD.

Brenton, A. (1994), *The Greening of Machiavelli: The Evolution of International Environmental Politics*, London, Royal Institute of International Affairs/Earthscan.

Bretherton, C. (1996), 'Gender and environmental change: are women the key to safeguarding the planet?', in J. F. Vogler and M. F. Imber (eds), *The Environment and International Relations*, London, Routledge, pp. 99–119.

Breyman, S. (1993), 'Knowledge as power: ecology movements and global environmental problems', in Lipschutz and Conca (1993).

British Medical Association (1999), *Biotechnology Weapons and Humanity*, Amsterdam, Harwood Academic Publishers.

Brownlie, I. (1990), *Principles of Public International Law*, 4th edn, Oxford, Oxford University Press.

Bud, R. (1993), *The Uses of Life: A History of Biotechnology*, Cambridge, Cambridge University Press.

Burton, J. (1972) *World Society*, Cambridge, Cambridge University Press.

Burton, J. (1999), 'The conjunction of competition and collaboration in international business', in R. Mudambi and M. Ricketts (eds), *The Organisation of the Firm: International Business Perspectives*, London, Routledge, pp. 102–25.

Busch, L. and A. Juska (1997), 'Beyond political economy: actor-networks and the globalization of agriculture', *Review of International Political Economy*, 4, 4: 688–708.

Busch, L., W. B. Lacy, J. Burkhardt and L. R. Lacy (1991), *Plants, Power, and Profit: Social, Economic, and Ethical Consequences of the New Biotechnologies*, Oxford, Basil Blackwell.

Business Week (1992), 'Biotech: America's dream machine', 2 March, 52–5.

Butler, L. J. (1996), 'Plant breeders' rights in the US: update of a 1983 study', in J. van Wijk and W. Jaffe (eds), *Intellectual Property Rights and Agriculture in Develop-*

ing Countries, University of Amsterdam.

Butler, L. J. and B. W. Marion (1985), 'The impacts of patent protection on the US seed industry and public plant breeding', University of Wisconsin.

Cadrin, C. and M. S. Golbus (1993), 'Fetal tissue sampling: indications, techniques, complications and experience with sampling of fetal skin, liver and muscle', *Western Journal of Medicine*, 159, 3: 269–72.

Caine, C. G. (1999), 'Technology outpaces restraints', *International Herald Tribune*, 16 April, p. 8.

Callon, M. (1991), 'Techno-economic networks and irreversibility', in Law (1991), pp. 132–61.

Callon, M. (1993), 'Variety and irreversibility in networks of technique conception and adoption', in D. Foray and C. Freeman (eds), *Technology and the Wealth of Nations*, London, Pinter, pp. 232–68.

Cambrosio, A. and P. Keating (1995), *Exquisite Specificity: The Monoclonal Antibody Revolution*, Oxford, Oxford University Press.

Campanella, M. L. (1995), 'The effects of globalization and turbulence on policy-making processes', in J. Drew (ed.), *Readings in International Enterprise*, London, Routledge, pp. 15–25.

Cantley, M. F. and J. Dreux de Nettancourt (1992), 'Biotechnology research and policy in the European Community: the first decade and a half', *Federation of European Microbiological Societies Letters*, 100, 25–32.

Carlsson, B. and S. Jacobsson (1993), 'Technological systems and economic performance: the diffusion of factory automation in Sweden', in Foray and Freeman (1993), pp. 77–92.

Carson, R. (1963), *Silent Spring*, London, Hamish Hamilton.

Carter, A. and D. Omand (1996), 'Countering the proliferation risks: adapting the alliance to the new security environment', *NATO Review*, 5: 10–15.

Carter, G. and G. S. Pearson (1996), 'North Atlantic chemical and biological research collaboration: 1916–1995', *Journal of Strategic Studies*, 19, 1: 74–103.

Carus, S. W. (1997), 'The threat of bioterrorism', *Strategic Forum*, 127.

Cauley, J. A., F. L. Lucas, L. H. Kuller, K. Stone, W. Browner, and S. R. Cummings, (1999), 'Elevated serum estradiol and testosterone concentrations are associated with a high risk for breast cancer', *Annals of Internal Medicine*, 130, 4: 270–8.

CBD Secretariat (1997), 'Background document on existing international agreements related to biosafety', UNEP/CBD/BSWG/2/3, Montreal.

CBD Secretariat (1999), 'Overview and annotated text of the protocol on biosafety', UNEP/CBD/BSWG/6/8, Montreal.

Center for Surrogate Parenting (undated), http://www.creatingfamilies.com

Central Intelligence Agency (1996), *The Chemical and Biological Weapons Threat*, CIA.

Cerny, P. G. (1996), 'What next for the nation-state?', in E. Kofman and G. Youngs (eds), *Globalization: Theory and Practice*, London, Pinter, pp. 123–37.

Chan, C. K. (1992), 'Eugenics on the rise: a report from Singapore', in R. F. Chadwick (ed.), *Ethics, Reproduction and Genetic Control*, London, Routledge, pp. 164–71.

Chemistry and Industry (1992), 'Watching out for Japan', No.11, June, 397.

Chevrier, M. (1996) 'Strengthening the Biological Weapons Convention', *Disarmament Diplomacy*.

Chimni, B. S. (1993), 'The philosophy of patents: strong regime unjustified', *Journal of Scientific and Industrial Research*, 52, April: 234–9.

Chodorow, N. (1978), *The Reproduction of Mothering*, Berkeley, University of California Press.

Christensen, G., Y. B. Wang and K. R. Chien (1997), 'Physiological assessment of complex cardiac phenotypes in genetically engineered mice', *American Journal of Physiology: Heart and Circulatory Physiology*, 41,6: h2513–h2524.

Christian Science Monitor (1999a), 'New genes meet a wary market', 8 December.

Christian Science Monitor (1999b), 'US poised for a biotech food fight', 17 November.

Coghlan, A. (1995), 'Sweeping patent shocks gene therapists', *New Scientist*, No.1971, April: 4.

Cohen, W. (1997), 'Proliferation: threat and response', Department of Defense, Washington, DC. Text available on World-wide Website, http://www.defenselink.mil/pulors/prolif97/index.html

Cole, L. A. (1997), *The Eleventh Plague: The Politics of Biological and Chemical Warfare*, New York, W. H. Freeman and Co.

COM (94) 319 Final (1994), Communication from the Commission to the Council to the European Parliament, Economic and Social Committee and the Committee of the Regions. An industrial competitiveness policy for the European Union.

Commission of the European Communities (CEC) (1997a), 'Biotechnology – five year assessment', Report EUR 17591.

Commission of the European Communities (CEC) (1997b), 'Co-operation with third countries and international organizations – five year assessment', Report EUR 17597.

Commission of the European Communities (CEC) (1997c), 'Biomedicine and health – five year assessment', Report EUR 17592.

Commission of the European Communities (CEC) (1999), 'Press Release: Second Conference of the Biotechnology and Finance Forum', Lyon, 26–29 March, DG XII Press Office, June 1999, at http://europa.eu.int/comm/dg12/press/1999/pr2503en.html

Committee on Japan (1992), *US–Japan Technology Linkages in Biotechnology: Challenges for the 1990s*, Washington, DC, National Academy Press.

Compeerapap, J. (1997), 'The Thai debate on biotechnology and regulations', *Biotechnology and Development Monitor*, no.32, September, 13–15.

Connolly, W. E. (1989), 'Identity and difference in global politics', in J. Der Derian and M. J. Shapiro (eds), *International/Intertextual Relations: Postmodern Readings of World Politics*, Lexington, MA, Lexington Books.

Consumers' Association (1997), 'Gene cuisine – a consumer agenda for genetically modified foods', *Policy Report*, London Consumers' Association.

Correa, C. M. (1992), 'Biological resources and intellectual property rights', *EIPR*, 14, 5: 154–7.

Correa, C. M. (1994), 'The GATT agreement on Trade-Related Aspects of Intellectual Property Rights: New standards for patent protection', *EIPR*, 16, 8: 327–35.

Cox, R. (1987), *Production, Power, and World Order*, New York, Columbia University Press.

Crawford, J., J. Dugard, P. Heyman, M. Meselson and J. P. Robinson (1998), 'A Draft Convention to prohibit biological and chemical weapons under international criminal law', *The CBW Conventions Bulletin*, issue number 42, December.

Croft, S. (1996), *Strategies of Arms Control: A History and Typology*, Manchester, Manchester University Press.

Daily Telegraph (1999), 'Monsanto in $50 bn merger with Pharmacia & Upjohn', 21 December.

Dalton, D. H. and M. G. Serapio (Jr) (1995), 'Globalizing industrial research and development', US Department of Commerce, Office of Technology Policy, Asia-Pacific Technology Program, October.

Daly, P. (1985), *The Biotechnology Business: A Strategic Analysis*, London, Frances Pinter, and Totowa, NJ, Rowman & Allanheld.

Dando, M. R. (1994), *Biological Warfare in the 21st century*, London, Brasseys.

Dando, M. R. (1999), 'The impact of the development of modern biology and medicine on the evolution of offensive biological warfare programs in the twentieth century', *Defense Analysis*, 15, 1: 43–62.

DaSilva E. J. and M. Taylor (1998), 'Island Communities and Biotechnology', *Electronic Journal of Biotechnology*, 1, 1: 1–10.

Davey, J. (1998), 'IT focus on pharmaceuticals', *Pharmaceutical Technology Europe*, 10, 10: 91–2.

Daza, C. (1998), 'Scientific research and training in biotechnology in Latin America and the Caribbean: the UNU/BIOLAC experience', *Electronic Journal of Biotechnology*, 1, 2: 18–24.

De Almeida, P. R. (1995), 'The political economy of intellectual property protection: Technological protectionism and transfer of revenue among nations', *International Journal of Technology Management*, 10, 2/3: 215–29.

De Kathen, A. (1999), 'Transgenic crops in developing countries', *Texte* 58, 99, Berlin, Umweltbundesamt.

DelGiudice, G., M. Pizza and R. Rappuoli (1998), 'Molecular aspects of medicine: A review journal for physicians and biomedical scientists', *Molecular Aspects of Medicine*, 19, 1: 1–69.

DG XII (1996), 'Biotechnology R and D in Europe: national files', EUR 17459 EN.

DG XII (1997), 'Report of the biotechnology entrepreneurship workshop', Amsterdam, 27 June.

DG XII (1998), 'Annual Report of EU R and D activities', Luxembourg, Office of Official Publications of the European Communities.

Di Berardino, M. A. (1999), 'Cloning: past, present, and the exciting future', published at http://www.faseb.org/opar/cloning

Dorabjee, S., C. E. Lumley and S. Cartwright (1998), 'Culture, innovation and successful development of new medicines – an exploratory study of the pharmaceutical industry', *Leadership and Organization Development Journal*, 19, 4: 199–210.

Dosi, G. (1984), 'Technological paradigms and technological trajectories. The determinants and directions of technical change and the transformation of the economy', in C. Freeman (ed.), *Long Waves in the World Economy*, London, Pinter, pp. 78–101.

Dosi, G. (1988), 'Sources, procedures and microeconomic effects of innovation', *Journal of Economic Literature*, 26: 1120–71.

Dosi, G. *et al.* (eds) (1988), *Technical Change and Economic Theory*, London, Pinter.

Douglas, M. and A. Wildavsky (1982), *Risk and Culture: An Essay on the Selection of Technical and Environmental Dangers*, Berkeley, CA, University of California Press.

Douglass, J. (1990), 'Beyond nuclear war', *The Journal of Social, Political and Economic Studies*, 15: 141–56.

Dower, N. (ed.) (1990), *Ethics and Environmental Responsibility*, Aldershot, Edgar Elgar.

DR Report (1996), 'Outlook for biopharmaceutical technologies to 2005', San Francisco.

DR Report (1997), 'The biopharmaceutical industry in transition: prospects for growth', San Francisco.

Dryzek, J. 1997, *The Politics of the Earth: Environmental Discourses*, Oxford, Oxford University Press.

DTI (Department of Trade and Industry) (1999), *Bioguide*, London, Stationery Office.

Dunning, J. (1993), *The Globalisation of Business: The Challenge of the 1990s*, London and New York, Routledge.

Dunning, J. and J. A. Cantwell (1991), 'MNEs, technology and the competitiveness of European industries', in G. R. Faulhaber and G. Tamburini (eds), *European Economic Integration*, Netherlands, Kluwer, pp. 117–48.

Durrani, M. (1996), 'Never mind the research, how's the cash flow?', *Physics World*, 9, 11: 23–6.

Dyson, A. and J. Harris (eds) (1994), *Ethics and Biotechnology*, London.

Eaglstein, W. H. and V. Falanga (1997), 'Tissue engineering and the development of Apligraf(R), a human skin equivalent', *Clinical Therapeutics*, 19, 5: 894–905.

Eckersley, R. (1992), *Environmentalism and Political Theory: Toward an Ecocentric Approach*, London, UCL Press.

Ecologist (1993), 'Whose common future? Reclaiming the commons', London, Earthscan.

The Economist (1991), 'Promises, promises, promises', 5 October, 89–90.

The Economist (1995), 'Biotechnology in Japan: alien culture', 18 November, 115.

Egziabher, T. B. G. (1999), 'Abdication of responsibility for biosafety in the name of free trade', Third World Network (TWN), http://www.twnside.org.sg/souths/twn/title/abdicate-cn.html

Elkington, J. (1985), *The Gene Factory: Inside the Biotechnology Business*, London, Century Publishing.

Elliot, L. (1998), *The Global Politics of the Environment*, London, Macmillan.

ENDS Daily (2000), 'Labeling of foods and food ingredients within the European Union that have been genetically modified or have been produced from genetically modified organisms', 12 January.

Engel, J. R. and J. G. Engel (eds) (1990), *Ethics of Environment and Development: Global Challenge and International Response*, Tucson, AZ, University of Arizona Press.

ESRC (1999), 'The politics of GM food: risk, science and public trust', Global environmental change programme, Special Briefing No. 5, University of Sussex.

Etherton, T. D., and D. E. Bauman (1998), 'Biology of somatotropin in growth and lactation of domestic animals', *Physiological Reviews*, 78, 3: 745–61.

European Council (1990), 'Council Directive of 23 April 1990 on the Contained Use of Genetically Modified Micro-Organisms' and 'Council Directive of 23 April 1990 on the Deliberate Release into the Environment of Genetically Modified Organisms', *Official Journal of the European Communities*, L117, 8 May: 1–27.

European Federation of Biotechnology (1996), 'Patenting in Biotechnology', Briefing Paper no.1.

Evers, S. (1997), 'US cities will be aided against toxic attack', *Jane's Defence Weekly*, 23 April, 6.

Fairhall, D. (1998), 'Russia continues to develop new germ weapons', *Guardian*, 26 February, 12.

Falk, R. (1975), *A Study of Future Worlds*, New York, Free Press.

Falk, R. (1990), 'Inhibiting reliance on biological weaponry: The role and relevance of international law', in Susan Wright (ed.), *Preventing a Biological Arms Race*, Cambridge, MA, MIT Press, pp. 241–66.

Falkenrath, R. (1998) 'Confronting nuclear, biological and chemical terrorism', *Survival*, 40, 3: 43–65.

Falkenrath, R., R. Newman and B. Thayer (1998), *America's Achilles' Heel: Nuclear, Biological, and Chemical Terrorism and Covert Attack*, Cambridge, MA, MIT Press.

Falkner, R. (1999), 'Frankenstein or benign? Genetics and trade', *The World Today*, 55, 7: 24–6.

Falkner, R. (2000a), 'The Biosafety Protocol: Trade and development policy implications', in *International Affairs*, April.

Falkner, R. (2000b), 'Business conflict and American foreign environmental policy', in P. G. Harris (ed.), *The Environment and American Foreign Policy*, Washington, DC, Georgetown University Press.

Farquhar, D. (1996), *The Other Machine: Discourse and Reproductive Technologies*, London, Routledge.

Farrands, C. (1997), 'Interpretations of the diffusion and absorption of technology: change in the global political economy', in Talalay, Farrands and Tooze (eds) (1997), pp. 75–89.

Farrell, M. (1996), 'European technology policy – the making of a technological community', European Dossier Series, no.37, London, North London University Press.

Federal Co-ordinating Council for Science, Engineering and Technology (FCCSET) (1992), 'Biotechnology for the 21st Century: Realizing the Promise', sourced, April 1999, at http://www.nal.usda.gov/bic/Federal_Biotech/biotech94.fccset.html

Federation of American Scientists (1995), 'Proposals for technological cooperation to implement Article X of the Biological Weapons Convention', 4th Workshop of the Pugwash Study Group on the Implementation of the Chemical and Biological Weapons Conventions: Strengthening the Biological Weapons Convention, 2–3 December, Geneva, Switzerland.

Financial Times (1999a), 'American shoppers taste 'benefits' of genetically altered food products', 19 February.

Financial Times (1999b), 'Funds threat to US biotech industry', 8 March.

Financial Times (1999c), 'Modified products banned by three fast food chains', 8 March.

Financial Times (1999d), 'EU rules "threaten" US biotechnology exports', 16 March.

Financial Times (1999e), 'Biotech companies stake out Brazilian battleground', 28 July.

Financial Times (1999f), 'Biotechnology: minister underlines support', 7 September.

Financial Times (2000), 'Novartis targets US mergers', 19 January.

Finder, J. (1986), 'Biological warfare, genetic engineering, and the treaty that failed', *The Washington Quarterly*, Spring 1986.

Fischhoff, B., S. Lichtenstein and P. Slovic (1980), 'Approaches to Acceptable Risk: A Critical Guide', Oak Ridge National Laboratory and the US Nuclear Regulatory Commission.

Foray, D. and C. Freeman (eds) (1993), *Technology and the Wealth of Nations*, London, Pinter.

Fox, W. (1990), *Towards a Transpersonal Ecology: Developing New Foundations for Environmentalism*, Boston, MA, Shambhala.

Franco, C. M. M. and N. C. McClure (1998), 'Isolation of microorganisms for biotechnological application', *Journal of Microbiology and Biotechnology*, 8, 2: 101–10.

Franz, D. R. *et al.* (1997), 'Clinical recognition and management of patients exposed to biological warfare agents', *Journal of the American Medical Association*, 278, 5: 399–411.

Freedman, L. and E. Karsh (eds) (1993), *The Gulf Conflict, 1990–1991*, Princeton, NJ, Princeton University Press.

Freeman, C. (1989), 'The diffusion of biotechnology through the economy: the timescale', in OECD, *Biotechnology: Economic and Wider Impacts*, Paris, OECD.

Freeman, C. (ed.) (1990), *The Economics of Innovation*, Aldershot, Edgar Elgar.

Freeman, C. and C. Perez (1988), 'Structural crises of adjustment, business cycles and investment behaviour', in Dosi *et al.* (eds) (1988), pp. 38–66.

Fritsch, M. (1995), 'The Market, market failure and the evaluation of technology-promoting programs', in G. Becher and S. Kaufmann (eds), *Evaluation of Technology Policy Programmes in Germany*, Netherlands, Kluwer, pp. 124–213.

Fukuyama, F. (1992), *The End of History and the Last Man*, London, Penguin.

Furtado, S. and O. Suchowerksky (1995), 'Huntingtons disease: recent advances in diagnosis and management', *Canadian Journal of Neurological Sciences*, 22, 1: 5–12.

Gaia Foundation and GRAIN (1998a) 'TRIPs versus CBD', *Global Trade and Biodiversity in Conflict*, April, London, The Gaia Foundation.

Gaia Foundation and GRAIN (1998b), 'Intellectual property rights and biodiversity: The economic myths', *Global Trade and Biodiversity in Conflict*, October, London, The Gaia Foundation.

Galhardi, R. M. A. A. (1994), *Small High Technology Firms in Developing Countries: The Case of Biotechnology*, Aldershot, Averbury Press.

GATT (1994), 'The results of the Uruguay Round of Multilateral Trade Negotiations: the legal texts', Geneva, GATT Secretariat.

Gaull, G. F. (1997), 'Biotechnology Regulation in America and Europe', Institute of Economic Affairs, Environment Working Paper No. 2, London, IEA.

Geissler, E. (ed.) (1986), *Biological and Toxin Weapons Today*, Oxford, Oxford University Press and SIPRI.

Gellissen, G. and C. P. Hollenberg (1997), 'Application of yeasts in gene expression studies: a comparison of *Saccharomyces cerevisiae*, *Hansenula polymorpha* and *Kluyveromyces lactis* – a review', *Gene*, 190, 1: 87–97.

Gendel, S. M., A. D. Kline, D. M. Warren and F. Yates (eds) (1990), *Agricultural Bioethics: Implications of Agricultural Biotechnology*, Ames, Iowa State University Press.

Genentech (1998), 'Monoclonal antibody technology', published at http://www.accessexcellence.org/ab/gg/monoclonal.html

The Genetics Forum (1993), *Genetic Engineering*, November.

George, S. (1990), *Ill Fares the Land*, London, Penguin.

Gilbert School (1999), 'Biotechnology: past, present and future', published at http://www.gilbertschool.org/students/larosa/biotechnology/history.html

Gill, P., A. J. Jeffreys and D. J. Werrett (1985), 'Forensic application of DNA fingerprints', *Nature*, 318, 6046: 577–9.

Gill, S (ed.) (1993), *Gramsci, Historical Materialism and International Relations*, Cambridge, Cambridge University Press.

Gilpin, R. (1987), *The Political Economy of International Relations*, Princeton, NJ, Princeton University Press.

Ginzburg, L. R. (ed.) (1991), *Assessing Ecological Risks of Biotechnology*, Stoneham, MA, Butterworth-Heinemann.

Glasner, P. and H. Rothman (eds) (1998), *Genetic Imaginations: Ethical, Legal and Social Issues in Human Genome Research*, Aldershot, Ashgate.

Glickman, T. S. and M. Gough (eds) (1990), *Readings in Risk*, Washington, DC, Resources for the Future.

Goldblat, J. (1971), 'The problem of chemical and biological warfare: A study of the historical, technical, military, legal and political aspects of CBW, and possible disarmament measures: Volume IV CB Disarmament Negotiations, 1920–1970', SIPRI, Stockholm, Almqvist and Wiksell.

Goodin, R. E. (1992), *Green Political Theory*, Cambridge, Polity Press.

Gottschalk, U. and S. Chan (1998), 'Somatic gene therapy – present situation and future perspective', *Arzneimittel Forschung Drug Research*, 48, 11: 1111–20.

Grace, E. S. (1997), *Biotechnology Unzipped: Promises and Realities*, Joseph Henry Press.

Granville, B. (1999), 'Bananas, beef and biotechnology', *The World Today*, 55, 4: 4–5.

Grasius, M. G., L. Iyengar and C. Venkobachar (1997), 'Anaerobic biotechnology for the treatment of wastewaters: a review', *Journal of Scientific and Industrial Research*, 56, 7: 385–97.

Grobstein, C. (1979), *A Double Image of the Double Helix: The Recombinant DNA Debate*, San Francisco, W.H. Freeman and Co.

Guardian (1999a), 'GM foods overtake BSE as top safety concern, says survey', 4 September.

Guardian (1999b), 'Blair condemns GM food "nonsense"', 25 September.

Guardian (1999c), 'GM labelling bill alarms US agencies', 1 November.

Guzzini, S. (1993), 'Structural power: The limits of neorealist analysis', *International Organization*, 47, 3: 443–78.

Haas, P. M. (1990), *Saving the Mediterranean: The Politics of International Environmental Cooperation*, New York, Columbia University Press.

Haas, P. M. (1991), 'Policy responses to stratospheric ozone depletion', *Global Environmental Change*, 1, 3, June.

Haas, P. M. (ed) (1992), 'Knowledge, Power, and International Policy Coordination', special edition, *International Organization*, 46, 1.

Halliday, F. (1994), *Rethinking International Relations*, London, Macmillan.

Hanson, P. P. (ed.) (1986), *Environmental Ethics: Philosophical and Policy Perspectives*, Burnaby, BC, Institute for the Humanities / Simon Fraser University.

Haraway, D. (1991), *Simians, Cyborgs and Women: The Reinvention of Nature*, London, Free Association Books.

Harding, K., E. E. Benson and K. Clacher (1997), 'Plant conservation biotechnology: an overview', *Agro Food Industry Hi-Tech*, 8, 3: 24–9.

Harding, S. (1991), *Whose Science? Whose Knowledge?*, Milton Keynes, Open University Press.

Hare, R. (1970), *Birth of Penicillin and the Disarming of Microbes*, London, George Allen & Unwin.

Harris, J. (1992), *Wonderwoman and Superman: The Ethics of Human Biotechnology*, Oxford, Oxford University Press.

Harris, S. H. (1994), *Factories of Death: Japanese Biological Warfare 1932–45 and the American Cover-up*, London, Routledge.

Harsanyi, Z. (8–9 June 1981), 'Statement by Zsolt Harsanyi: Commercialization of Academic Biomedical Research', Hearing before the Subcommittee on Investigations and Oversight and the Subcommittee on Science, Research, and Technology of the Committee on Science, Research, and Technology, US House of Representatives, ninety-seventh Congress, no.46, USGPO, Washington, DC.

Hart, S. E. and L. M. Wax (1999), 'Review and future prospectus on the impacts of herbicide resistant maize on weed management', *Maydica*, 44, 1: 25–36.

Hayward, S. (1997), 'A political economy approach to labour markets in knowledge-intensive industries: the case of biotechnology', in Talalay, Farrands and Tooze (eds) (1997), pp. 126–38.

Hayward, S. (1998), 'Towards a political economy of biotechnology development: a sectoral analysis of Europe', *New Political Economy*, 3, 1: 79–101.

Held, D., A. McGrew, D. Goldblatt and J. Perraton (1999), *Global Transformations*, Cambridge, Polity Press.

Hileman, B. (1999), 'UK moratorium on biotech crops urged', *Chemical and Engineering News*, 24 May, 7.

Hobbelink, H. (1991), *Biotechnology and the Future of World Agriculture*, London, Zed Books.

Holland, A. and A. Johnson (eds) (1998), *Animal Biotechnology and Ethics*, London, Chapman & Hall.

Holland, J. (1999), 'DuPont to pay $7.7 billion for agribusiness giant', *San Diego Union Tribune*, 16 March, C-2.

Holland, S. (1993), *The European Imperative: Economic and Social Cohesion in the 1990s*, London, Spokesman Press.

Hopgood, S. (1998), *American Foreign Environmetal Policy and the Power of the State*, Oxford, Oxford University Press.

Howey, D. C., S. E. Fineberg, P. A. Nolen, M. I. Stone, R. G. Gibson, N. S. Fineberg and J. A. Galloway (1982), 'The therapeutic efficacy of human insulin (recombinant DNA) in patients with insulin-dependent diabetes-mellitus: a comparative study with purified porcine insulin', *Diabetes Care*, 5, S2: 73–7.

Human Rights Watch/Asia (undated), www.american.edu/projects/mandala/ted/body.html

Humbert, M. (1994), 'Strategic industrial policies in a global industrial system', *Review of International Political Economy*, 1, 3: 445–63.

Hunger, I. (1996), 'Article V: confidence building measures', in G. S. Pearson and M. R. Dando (eds), *Strengthening the Biological Weapons Convention: Key Points for the Fourth Review Conference*, Bradford, University of Bradford, pp. 77–92.

Hurrell, A. and B. Kingsbury (eds) (1992), *The International Politics of the Environment: Actors, Interests, and Institutions*, Oxford, Clarendon Press.

Hynes, H. P. (1989), *The Recurring Silent Spring*, Oxford, Pergamon Press.

Inglehart, R. (1997), *Modernization and Postmodernization*, Princeton, NJ, Princeton University Press.

Institute for Science in Society (1999), 'Special Safety Concerns of Transgenic Agriculture and Related Issues' , a briefing paper for Minister of State for the Environment, the Rt. Hon. Michael Meacher.

IPTS (1997), 'A prospective analysis of European pharmaceutical research development and innovation', IPTS Working Report.

Isken, S. and J. A. M. deBont (1998), 'Bacteria tolerant to organic solvents', *Extremophiles*, 2, 3: 229–38.

James, C. (1999), 'Global status of commercialized transgenic crops: 1999', *ISAAA Briefs* No. 12: Preview, Ithaca, NY, ISAAA.

Jane's Defence Weekly, 26 August 1998, 8. Cited in (1998), 'Science and technology scan', *Trust and Verify*, 82: 5.

Jeffreys, A. J., W. Wilson and S. L. Thein (1985), 'Individual specific fingerprinting of human DNA', *Nature*, 316, 6021: 76–9.

Jelsma, J. (1995), 'Military implications of biotechnology', in M. Fransman, G. Junne, and A. Roobeck (eds), *The Biotechnology Revolution*, Oxford, Blackwell.

Jenkins, N., R. B. Parekh and D. C. James (1996), 'Getting the glycosylation right: implications for the biotechnology industry', *Nature Biotechnology*, 14, 8: 975–81.

Jensen, B. B. (1998), 'The impact of feed additives on the microbial ecology of the gut in young pigs', *Journal of Animal and Feed Sciences*, 17, s1, 45–64.

Joint Technical Warfare Committee (1945), 'Potentialities of weapons of war during the next ten years', Chief of Staff Committee, T.W.C. (45) 42, 12 November.

Jones, R. J. B. (1995), *Globalisation and Interdependence in the International Political Economy: Rhetoric and Reality*, London and New York, Pinter Publishers.

Jordan, M. A., S. McGinness and C. V. Phillips (1996), 'Acidophilic bacteria: their potential mining and environmental applications', *Minerals Engineering*, 9, 2: 169–81.

Joseph, R. B and J. F. Reichart (1996), 'Deterrence and defense in a nuclear, biological, and chemical environment', *Comparative Strategy*, 15: 59–80.

Juma, C. (1989), *The Gene Hunters: Biotechnology and the Scramble for Seeds*, London, Zed Books.

Kawamura, K. (1998), 'International harmonization of regulation in Japan – what does the future hold?', *Pharmaceutical Technology Europe*, 10: 10, 38–47.

Kealey, D.A. (1990), 'Towards an integral ecological ethic', in *Revisioning Environmental Ethics*, Albany, NY, State University of New York Press.

Keeble, S. (1994), 'Infertility, feminism and the new technologies', *Fabian Pamphlet*, 566, London, Fabian Society.

Kenney, M. (1986) *Biotechnology: The University-Industrial Complex*, New Haven, CO, Yale University Press.

Keohane, R. O. (1984), *After Hegemony: Cooperation and Discord in World Political Economy*, Princeton, NJ, Princeton University Press.

Keohane, R. O. (1993), 'The analysis of international regimes: a European–American research programme', in V. Rittberger and P. Mayer (eds), *Regime Theory and International Relations*, Oxford, Clarendon Press, pp. 23–48.

Keohane, R. O. and J. S. Nye (1977), *Power and Interdependence*, Boston, Toronto, Little, Brown and Co.

Khalid, F. (1992), *Islam and Ecology*, London, Cassell.

Kinoshita, J. (1993), 'Is Japan a boon or a burden to US industry's leadership?', *Science*, 259, 29 January, 596–8.

Kinter, W. R. and H. Sicherman (1975), *Technology and International Politics: The Crisis of Wishing*, Lexington, MA, Lexington Books.

Kipping, M. (1997), 'European industrial policy in a competitive global economy', in S. Stavridis, E. Mossialos, R. Morgan and H. Machin (eds), *New Challenges to the European Union: Policies and Policy Making*, Dartmouth, Dartmouth Publishing House, pp. 489–518.

Kloppenburg, J. R. (1988), *First the Seed: the Political Economy of Plant Biotechnology*, Cambridge, Cambridge University Press.

Ko, Y. (1992), 'An economic analysis of biotechnology patent protection', *The Yale Law Journal*, 102, 747: 777–804.

Köhler, G. and C. Milstein (1975), 'Continuous cultures of fused cells secreting antibody of redefined specificity', *Nature*, 256: 495–7.

Kolakowski, L. (1988), *Metaphysical Horror*, Oxford, Basil Blackwell.

Krimsky, S. (1982), *Genetic Alchemy: The Social History of the Recombinant DNA Controversy*, Cambridge, MA, MIT Press.

Kuhn, T. (1970), *The Structure of Scientific Revolutions*, 2nd edn, Chigaco and London, University of Chicago Press.

Kupper, T. (1999), 'Bioprospecting goes to the ends of the earth in search of wonder drugs', *San Diego Union-Tribune*, 10 January, I-1.

Kyriakou D. and D. Gilson (1998), 'Biotechnology and healthcare: consumer related aspects', *The IPTS Report*, 30: 20–6.

Lacey, E. J. (1994), 'Tackling the biological weapons threat: the next proliferation challenge', *The Washington Quarterly*, 17, 4: 53–64.

LaFee, S. (1999), 'Smart money: in science today, the question is: who will own your next great idea?', *San Diego Union Tribune*, 24 March, E-1.

Larson, A. (1999), 'Biotechnology: finding a practical approach to a promising technology', *Economic Perspectives* (Electronic Journal of the US Department of State), 4, 4, October, http://www.usia.gov/journals/ites/1099/bio-toc.html

Larson, R. J. and R. P. Kadlec (1996), 'Biological Warfare – its technical aspects', *Army Chemical Review*, July, 27–32.

Latour, B. (1991), 'Technology is society made durable', in J. Law, (ed.) (1991), pp. 103–31.

Laughlin, L. L. Jr (1977), *US Army Activity in the US Biological Warfare Programs*, Vol II, Dept. of the Army, Washington, DC.

Law, J. (ed.) (1991), *A Sociology of Monsters*, London and New York, Routledge.

Lawton, T. (1997), *Technology and the New Diplomacy*, Aldershot, Avebury.

Lee, L. E. J., P. Chin and D. D. Mosser (1998), 'Biotechnology and the Internet', *Biotechnology Advances*, 16: 949–60.

Lemieux, B., A. Aharoni and M. Schena (1998), 'Overview of DNA chip technology', *Molecular Breeding*, 4, 4: 277–89.

Levidow, L. (1999), 'Regulating BT maize in the United States and Europe: a scientific-cultural comparison', *Environment*, 41, 10: 11–22.

Lewin, B., P. Siliciano and M. Klotz (1997), *Genes VI*, Oxford, Oxford University Press.

Lipschutz, R. D. and K. Conca (eds) (1993), *The State and Social Power in Global Environmental Politics*, New York, NY, Columbia University Press.

Lissens, W. and K. Sermon (1997), 'Preimplantation genetic diagnosis: current status and new developments', *Human Reproduction*, 12, 8: 1756–61.

Loi, P., S. Boyazoglu, M. Gallus, S. Ledda, S. Naitana, I. Wilmut, P. Cappai and S. Casu (1997), 'Embryo cloning in sheep: work in progress', *Theriogenology*, 48, 1: 1–10.

Lynn. M. (1993), 'Biotech grows on market', *Sunday Times*, 20 June, 6.

MacMillan, J. and A. Linklater (1995), *Boundaries in Question: New Directions in International Relations*, London, Pinter.

Magee, M. (1999), 'High Tech High, a school geared to industry, to open next year', *San Diego Union-Tribune*, 22 January, B-6.

Mantegazzini (1986), *The Environmental Risks From Biotechnology*, London, Frances Pinter.

Mark, H. F. L. and K. D. McGowan (1996), 'Issues in the genetic assessment of predispositions for familial breast and ovarian cancer', *Cytobios*, 87, 351: 229–35.

Martin, W. and L. A. Winters (eds) (1996), *The Uruguay Round and Developing Countries*, Cambridge, Cambridge University Press.

Mason C. M. and R. T. Harrison (1996), 'The UK clearing banks and the informal venture capital market', *International Journal of Bank Marketing*, 14, 1: 5–14.

Mason, C. and R. Harrison (1996), 'Developments in the promotion of informal venture capital in the UK', *International Journal of Entrepreneurial Behaviour and Research*, 2, 2: 6–33.

Mason, M. (1994), 'Elements of consensus: Europe's response to the Japanese automotive challenge', *Journal of Common Market Studies*, 32, 4: 435–53.

Mather, J. P. and P. E. Roberts (1998), *Introduction to Cell and Tissue Culture: Theory and Technique (Introductory Cell and Molecular Biology Techniques)*, Plenum Publishing Corporation.

McConnell, F. (1996), *The Biodiversity Convention: A Negotiating History* (A Personal Account of Negotiating the CBD – and After), London, Kluwer Law International.

McDougall, C. L and R. Hall (eds) (1996), *Intellectual Property Rights and Biodiversity Convention: The Impact of GATT*, London, Friends of the Earth.

McGraw, D. (2000), 'Negotiating the UN Biodiversity Convention: A Case Study in International Regime Formation', thesis submitted in partial fulfilment of the requirement for the Ph.D. degree in the Faculty of Economics of the University of London.

McGuire, S. (1999), 'Firms and governments in international trade', in B. Hocking and S. McGuire (eds), *Trade Politics: International, Domestic and Regional Perspectives*, London, Routledge, pp. 147–61.

McNally, K. (1995), 'Corporate venture capital: the financing of technology businesses', *International Journal of Entrepreneurial Behaviour and Research*, 1, 3: 9–43.

Mellor, M. (1997), *Feminism and Ecology*, Cambridge, Polity.

Merchant, C. (1982), *The Death of Nature: Women, Ecology and the Scientific Revolution*, London, Wildwood House.

Meselson, M. (1999), 'The problem of biological weapons', 11th Workshop of the Pugwash Study Group on the Implementation of the Chemical and Biological Weapons Conventions: Implications of the CWC Implementation for the BTWC Protocol Negotiations, 15–16 May, Noordwijk, The Netherlands. (Previously, Presentation given at the 1818th Stated Meeting of the American Academy of Arts and Sciences, 13 January, Cambridge).

Metcalfe, J. S. (1998), 'Science policy and technology policy in a competitive economy', *International Journal of Social Economics*, 24, 7/8: 723–40.

Metcalfe, S. (1997), 'Technology systems and technology policy in an evolutionary framework', in Archibugi and Michie (eds) (1997), pp. 268–96.

Michael, M. (1996), *Constructing Identities*, London, Sage.

Mies, M. (1986), *Patriarchy and Accumulation on a Global Scale*, London, Zed Books.

Mies, M. (1993), 'White man's dilemma: his search for what he has destroyed', in M. Mies and V. Shiva (1993), pp. 132–63.

Mies, M. and S. Shiva (1993), *Ecofeminism*, London, Zed Books.

Miller D. L. (1952), 'History of air force participation in biological warfare program: 1944–1951', Historical Study 194, Wright-Patterson Air Force Base, September (the 'Miller Report').

Miller, J. and W. Broad (1998), 'War games show up germ defences', *Guardian*, 28 April, 13.

Mitchell, G. R. (1997), 'The Global Context for US Technology Policy', US Department of Commerce, Office of Technology Policy.

MITI (1998), 'Annual Report on the Promotion of Science and Technology 1998' (Summary), Ministry of International Trade and Industry (MITI), Japan, sourced, April 1999, at http://www.sta.go.jp/policy/seisaku/nenjiho96/index.html

Mitter, S. (1994), What women demand of technology, *New Left Review*, 205, May/June, 100–10.

Molina, A. H. (1995), *Constituency Building in the European Union: The Case of ESPRIT*, Luxembourg, OOPEC.

Moodie, M. (1994), 'Bolstering compliance with the biological weapons convention: Prospects for the special conference', *Chemical Weapons Convention Bulletin*, Issue Number 25, September.

Morgan, D. (1999), 'Monsanto's stock rises on talk of merger with DuPont', *San Diego Union Tribune*, 4 March, C-2.

Morgan, R. A. and W. F. Anderson (1993), 'Human gene therapy', *Annual Review of Biochemistry*, 62: 191–217.

Morse, S. (1994), 'Vaccines for public health: can vaccines for peace help in the war against disease?', in E. Geissler and J. Woodall (eds), *Control of Dual-Use Agents: The Vaccines for Peace Programme*, New York, Oxford University Press.

Moser, I. (1995), 'Introduction: Mobilizing critical communities and discourses on modern biotechnology', in V. Shiva and I. Moser (eds), *Biopolitics: A Feminist and Ecological Reader on Biotechnology*, London, Zed Books, pp. 1–24.

Moss, N. (1970), *Men Who Play God: The Story of the Hydrogen Bomb*, London, Penguin.

Mossenlechner, K. (1999), 'Implications for the BTWC Protocol negotiations: the right mixture of compliance measures', 11th Workshop of the Pugwash Study Group on the Implementation of the Chemical and Biological Weapons Conventions: Implications of the CWC Implementation for the BTWC Protocol Negotiations, 15–16 May, Noordwijk, The Netherlands.

Mullis, K. B., F. Ferre and R. A. Gibbs (1994), *The Polymerase Chain Reaction*, Springer Verlag.

Munson, A. (1993), 'The United Nations Convention on biodiversity', in M. Grubb *et al.*, *The Earth Summit Agreements: A Guide and Assessment*, London, RIIA/Earthscan, pp. 75–84.

Murdoch, J. (1997), 'Towards a geography of heterogenous associations', *Progress in Human Geography*, 21, 3: 321–37.

Murphy, A. and J. Perrella (1993a), 'About biotechnology, overview and brief history', published at http://www.accessexcellence.org/AB/BC/Overview_and_Brief_History.html

Murphy, A. and J. Perrella (1993b), 'About biotechnology, overview and brief history', published at http://www.accessexcellence.org/ae/aepc/wwc/1993/history.html

Nathanson, V., M. Darvell and M. R. Dando (1999), *Biotechnology Weapons and Humanity*, Harwood Academic Publishers (for the British Medical Association).

National Science and Technology Council (NSTC) (1999), 'Bioinformatics in the 21st century', A Report to the Research Resources and Infrastructure Working Group Subcommittee on Biotechnology National Science and Technology Council White House Office of Science and Technology Policy, Bioinformatics Workshop 3–4 February, Krasnow Institute for Advanced Study George Mason University Fairfax, Virginia at http://www2.whitehouse.gov/wh/eop/ostp/nstc/html/bioinformaticsreport.html, June 1999.

National Science Foundation (1995), 'Asia's new high tech competitors', sourced, June 1998, from http://www.nsf.gov/sbe/srs/s4495/s4495039.html

National Science Foundation (1996), 'Asia's new high tech competitors: competitiveness in the market place', sourced, June 1998, from http:// www.nsf.gov/sbe/srs/s4495/conten2a.html

National Science Foundation (1998), 'Federal R and D funding by budget function – 1998 – An overview', sourced, June 1998, at www.nsf.gov/sbe/srs/nsf98301/overview.html

National Science Foundation (1998a), NSF/Tokyo Report – Supplemental Budget request of the Science and Technology Agency.

Nature Biotechnology (1998), 'Bioentrepreneurship I', Supplement, May, 1–64.

Nature Biotechnology (1999a), 'Bioentrepreneurship II', Supplement, February, BE1–BE40.

Nature Biotechnology (1999b), 'Bioentrepreneurship III', Supplement, May, BE1–BE37.

Nature Conservancy Council (1990), *Environmental Philosophy, A Bibliography*, Centre for Philosophy and Public Affairs, University of St. Andrews.

Nau, H. R. (1974), *National Politics and International Technology: Nuclear Reactor Development in Western Europe*, Baltimore, Johns Hopkins.

Nelson, R. R. and S. G. Winter (1987), 'In search of a useful theory of innovation', *Research Policy*, 6, 1: 36–76.

New York Times (1999), 'New trade threats for US farmers', 29 August.

Newsedge (1999a), 'UK Government: Sainsbury announces biotechnology clusters team', 19 April.

Newsedge (1999b), 'Europe biotech sector sees growth despite UK pain', 21 April.

Newsedge (1999c), 'Biotechnology in need of funding – and respectability', 16 April, sourced at www.newsedge.com April 1999.

Newsedge (1999d), 'British biotechnology firm increases losses but cites breakthroughs', 22 April.

Newsedge (1999e), 'Cuban biopharmaceuticals look for ticket to Europe', 22 April.

Ng, S. C. S., A. W. Pearson and D. F. Ball (1992), 'Strategies of biotechnology companies', *Technology Analysis and Strategic Management*, 4, 4: 351–61.

Nicolaides, P. (1992), 'EC industrial policy', *EIU European Trends*, 2: 53–7.

NISTEP (1997), *Science and Technology Indicators: 1997*, NISTEP Agency, Report no.50, sourced, May 1999, from http://www.nistep.go.jp/achiev/report50-e/loc.html

North Carolina's Bioscience Industry Forum (1998), 'A strategy for enhancing competitiveness', sourced, April 1999, at http://www.nciobiotech.org/comindex.html

Norton-Taylor, R. (1999), '£270m chemical weapon defence for British troops', *Guardian*, 22 July, 8.

NSF (1998b), 'Japan's science and technology policy: retooling for the future', sourced, April 1999, at http://www.nsf.gov/pubs/1998/int9812/int9812.txt

Nuffield Council on Bioethics (1999), 'Genetically modified crops. The ethical and social issues', London, The Nuffield Foundation http://www.nuffield.org/bioethics/publication/modifiedcrops/index.html

Nugent, N. and R. O'Donnell (1994), *The European Business Environment*, London, Macmillan.

Nuland, S. B. (1994), *How We Die*, London, Vintage.

OECD (1988), *Biotechnology and the Changing Role of Government*, Paris, OECD.

OECD (1989), *Biotechnology: Economic and Wider Impacts*, Paris, OECD.

OECD (1997), 'Foreign access to technology programmes', Report (97) 209 Paris, OECD.

OECD (1998), *Biotechnology for Clean Industrial Products and Processes: Towards Industrial Sustainability*, Paris, OECD.

Oelschlaeger, M. (1991), 'The alchemy of modernism: the transmutation of wilderness into nature', *The Idea of Wilderness: From Prehistory to the Age of Ecology*, New Haven, Conn. and London, Yale University Press.

Office of Technology Assessment (1993), 'Proliferation of Weapons of Mass Destruction: Assessing the Risks', OTA-ISC-559, August, United States Congress.

Office of Technology Policy (OTP) (1997), 'International plans, policies and investments in science and technology', April.

OksmanCaldentey, K. M. and R. Hiltunen (1996), 'Transgenic crops for improved pharmaceutical products', *Field Crops Research*, 45, 1–3: 57–69.

Old, R. W. and S. B. Primrose (1994), *Principles of Gene Manipulation: An Introduction to Genetic Engineering*, 5th edn, Oxford, Blackwells.

Omari, C. K. (1990), 'Traditional African land ethics', in Engel and Engel (1990).

Orsenigo, L. (1989), *The Emergence of Biotechnology: Institutions and Markets in Industrial Innovation*, London, Pinter.

Panopoulos, N. J., E. Hatziloukas and A. S. Afendra (1996), 'Transgenic crop resistance to bacteria', *Field Crops Research*, 45, 1–3: 85–97.

Parker, I. M. and P. Kareiva (1996), 'Assessing the risk of invasion for genetically engineered plants', *Biological Conservation*, 78, 1–2: 193–203.

Patel, T. (1997), *India Kidney Trade*, http://www.gurukul.ucc.american.edu/ted/kidney.html

Pauls, K. P., (1995), 'Plant biotechnology for crop improvement', *Biotechnology Advances*, 13, 4: 673–93.

Pearce, F. (1998), 'Return of the plague', *Guardian*, Online Section, 29 January, 3.

Pearson, G. S (1993), 'Prospects for chemical and biological arms control: the web of deterrence', *The Washington Quarterly*, Spring, 145–62.

Pearson, G. S. (1997), 'The complementary role of environmental and security biological control regimes in the 21st Century' *Journal of the American Medical Association (JAMA)*, 276, 5: 369–72.

Pearson, G. S. (1998a), 'Article X: some building blocks', Strengthening the Biological Weapons Convention, Briefing Paper Number 6, Bradford, University of Bradford, March.

Pearson, G. S. (1998b), 'Article X: further building blocks', Strengthening the Biological Weapons Convention, Briefing Paper Number 7, Bradford, University of Bradford, March.

Pearson, G. S. (1998c), 'Article X: pharmaceutical building blocks', Strengthening the Biological Weapons Convention, Briefing Paper Number 8, Bradford, University of Bradford, July.

Pearson, G. S. (1998d), 'The protocol to strengthen the BTWC: An integrated regime', *Politics and the Life Sciences*, 17, 2, September: 189–201.

Pearson, G. S., (1998e), 'Biological Weapons Protocol update', *Trust and Verify*, 82.

Pepperall, R. (1995), *The Post-Human Condition*, London, Intellect Books.

Perkins, W. A., R. W. McMullen and L. M. Vaughan (1958), 'Current status of anti-crop warfare capability, Appendix VI, Operational effectiveness of biological warfare', *Operations Research Group Study*, Nr. 21, Volume 13 (of 15 Volumes), 1 August, US Army Chemical Corps Operations Research Group, Army Chemical Center, Maryland.

Peterson, J. (1994), *High Technology and the Competition State: An Analysis of the Eureka Initiative*, London, Routledge.

Pettman, J. J. (1996), *Worlding Women: A Feminist International Politics*, Sydney, NSW, Allen & Unwin.

Pile, J. C. *et al.* (1998), 'Anthrax as a potential biological warfare agent', *Arch. Intern. Med.*, 158, 9, March: 429–34.

Pinkert, C. A. (1994), *Transgenic Animal Technology: A Laboratory Handbook*, Academic Press.

Powell, C. and A. Pearson (1995), 'The development and survival of SMEs in the pharmaceutical industry', High Technology Small Firms Conference, Manchester Business School, 18–19 September.

Pownall, I. (1998), 'An entrepreneurial focus to UK new technology based firm policies', *Science and Public Policy*, 25, 2: 117–33.

Prentis, S. (1984), *Biotechnology: A New Industrial Revolution*, London, Orbis.

Prevezer, M. and S. Toker (1996), 'The degree of integration in strategic alliances in biotechnology', *Technology Analysis and Strategic Management*, 8, 2: 117–33.

Primrose, S. B. (1987), *Modern Biotechnology*, Blackwell Scientific.

Ramey, T. S., M. J. Wimmer and R. M. Rocker (1999), 'GMOs are Dead', Deutsche Bank Alex. Brown, 21 May. http://www.biotech-info.net/Deutsche.pdf

Ranalli, P. and J. I. Cubero (1997), 'Bases for genetic improvement of grain legumes', *Field Crops Research*, 53: 69–82.

Ranger, R. (ed.) (1996), *The Devils Brews I: Chemical and Biological Weapons and their Delivery Systems*.

Raustiala, K. and D. G. Victor (1996), 'Biodiversity since Rio: The future of the Convention on Biodiversity', *Environment*, 38, 4, May: 17–45.

Redclift, M. and C. Sage (1998), 'Global environmental change and global inequality', *International Sociology*, 13, 4, December.

Reed, C. (1998a), '"New York anthrax attack" foiled', *Guardian*, 20 February, 1.

Reed, C. (1998b), 'Mystery clouds germ warfare "plot"', *Guardian*, 21 February, 15.

Regis, E. (1990), *Great Mambo Chicken and the Transhuman Condition: Science Slightly over the Edge*, London, Penguin.

Reich, R. (1991), *The Work of Nations*, London and New York, Simon & Schuster.

Reichman, J. H. (1995), 'Universal minimum standards of intellectual property under the TRIPS Component of the WTO agreement', *The International Lawyer*, 29, 2: 345–85.

Reuters (1997), 'Cuba claims biological raid', *Guardian*, 26 August, 8.

Reuters (1999), 'FDA to hold meetings on genetically engineered foods', 19 October.

Reuters (1999a), 'Thailand says no GM seed imports for now', 18 October.

Rich, B. (1994), *Mortgaging the Earth: The World Bank, Environmental Impoverishment and the Crisis of Development*, London, Earthscan.

Ring, P. S., S. A. Lenway and M. Govekar (1990), 'Management of the political imperative in international business', *Strategic Management Journal*, 11: 141–51.

Roberts, B. (1998), 'Export controls and biological weapons: New roles, new challenges', in R. M. Atlas (ed.), *Critical Reviews in Microbiology: Special Issue Biological Weapons*, 34, 3: 235–54.

Roberts, B. (ed.) (1993), *Biological Weapons: Weapons of the Future*, Washington, DC, CSIS Books.

Roberts, T. (1994), 'Broad claims for biotechnological inventions', *EIPR*, 16: 371–3.

Robertson, R. (1992), *Globalization: Social Theory and Global Culture*, London, Sage.

Rogers, P., S. Whitby and M. Dando (1999), 'Biological warfare against crops', *Scientific American*, 280, 6, June: 70–5.

Rohricht, P. (1999), 'Transgenic protein production – The technology and major players', *Biopharm, the Applied Technologies of Biopharmaceutical Development*, 12, 3: 46–9.

Ronit, K. (1997), 'Academia-industry-government relations in biotechnology: Private, professional and public dimensions of the new associations', *Science and Public Policy*, 24, 6: 421–33.

Rose, H. (1983), Hand, brain and heart: a feminist epistemology for the natural sciences, *Signs* 9, 1: 73–90.

Rosenau, J. N. (1990), *Turbulence in World Politics*, London, Harvester-Wheatsheaf.

Rosenau, J. N. (1993), 'Environmental Challenges in a Turbulent World', in Lipschutz and Conca (1993).

Rosenberg, N. (1982), *Inside the Black Box*, Cambridge, Cambridge University Press.

Rosenberg, N. (1994), *Exploring the Black Box*, Cambridge, Cambridge University Press.

Ruddick, S. (1980), 'Maternal Thinking', *Feminist Studies*, 6, 342–67.

Rufford, N. (1998), 'Britain funds biological war against heroin', *Sunday Times*, 28 June.

Ruggie, J. G. (1975), 'International responses to technology: concepts and trends', *International Organization*, 29, 3: 557–83.

RuibalMendieta, N. L. and F. A. Lints (1998), 'Novel and transgenic food crops: overview of scientific versus public perception', *Transgenic Research*, 7, 5: 379–86.

Rural Advancement Foundation International (RAFI) (1994), 'Pirating medicinal plants, COPs and robbers', *Occasional Papers*, 1, 4: 1–8.

Rural Advancement Foundation International (RAFI) (1997), 'The Life Industry 1997', *RAFI Communiqué*, November/December, 1–5.

Rural Advancement Foundation International (RAFI) (1998), 'Seed Industry Consolidation: Who Owns Whom?', *RAFI Communiqué*, July/August, 1–6.

Russell, A. (1988), *The Biotechnology Revolution: An International Perspective*, Sussex, New York, Wheatsheaf, St. Martin's Press.

Russell, A. (1995), 'Merging technological paradigms and the knowledge structure in the global political economy', *Science and Public Policy*, 22, 2: 106–16.

Russell, A. (1999a), 'Biotechnology as a technological paradigm in the knowledge structure', *Technology Analysis & Strategic Management*, II, 2: 235–54.

Russell, A. (1999b), 'Actor-networks, international political economy and risk in genetic manipulation', *New Genetics and Society*, 18, 2/3: 157–79.

Sagoff, M. (1991), 'On making nature safe for biotechnology', in Ginzburg (ed.) (1991), 341–65.

Sanchez, V. and C. Juma (eds) (1994) *Biodiplomacy: Genetic Resources and International Relations*, Nairobi, African Centre for Technology Studies.

Sandholtz, W. (1992), *High Tech Europe: The Politics of Collective Action*, Berkeley, California University Press.

Sands, P. (ed.) (1993), *The Greening of International Law*, London, Earthscan.

Sanhueza, C. (1996), 'The obligation to cooperate in technology transfer: Article X of the BWC', *UNIDIR Newsletter*, Number 33/96, Geneva, pp. 80–4.

Sasson, A. (1993), *Biotechnologies in Developing Countries: Present and Future*, Paris, UNESCO.

Saxonhouse, G. (1986), 'Industrial policy and factor markets: biotechnology in Japan and the United States', in H. Patrick (ed.), *Japan's High Technology Industries*, Seattle and London, University of Washington Press, University of Tokyo Press, pp. 97–135.

Sayler, G. S., J. Sanseverino and K. L. Davis (eds) (1997), *Biotechnology in the Sustainable Environment*, New York and London, Plenum Press.

Schnieke, A. E., A. J. Kind, W. A. Ritchie, K. Mycock, A. R. Scott, M. Ritchie, I. Wilmut, A. Colman and K. H. S. Campbell (1997), 'Human factor IX sheep produced by transfer of nuclei from transfected fetal fibroblasts', *Science*, 278, 5346: 2130–3.

Schwartz, H. (1994), *States versus Markets*, New York, St. Martin's Press.

Science and Technology Agency (STA) (1998a), 'Annual Report on the promotion of science and technology, 1998 (Summary)', sourced, June 1999, at http://www.sta.go.jp/policy/seisaku/nenjiho96/index.html

Science and Technology Agency (STA) (1998b), *Globalization and the Japanese economy*, sourced, May 1999, at www.sta.go.jp/policy

Sell, S. K. (1995), 'The origins of a trade-based approach to intellectual property protection: The role of industry associations', *Science Communication*, 17, 2: 163–85.

Sharansky, I. (1995), 'Policy analysis in a historical perspective', *Journal of Management History*, 1, 1: 47–58.

Sharp, M. (1989), 'The Community and new technologies', in J. Lodge (ed.), *The European Community and the Challenge of the Future*, London, Pinter, pp. 202–20.

Sharp, M. (1991), 'Pharmaceuticals and biotechnology: perspectives for the European industry', in C. Freeman, M. Sharp and W. Walker (eds), *Technology and the Future of Europe*, London, Pinter.

Shiva, V. (1993), *Monocultures of the Mind: Perspectives on Biodiversity and Biotechnology*, London, Zed Books.

Shiva, V. and I. Moser (eds) (1995), *Biopolitics: A Feminist and Ecological Reader on Biotechnology*, London, Zed Books.

Shue, H. (1992), 'The unavoidability of justice', in Hurrell and Kingsbury (eds) (1992), pp. 373–97.

Sklair, L. (1998), 'Globalization and the corporations: the case of the California *Fortune* Global 500', *International Journal of Urban and Regional Research*, 22, 2: 195–215.

Skolnikoff, E. (1993), *The Elusive Transformation: Science, Technology and the Evolution of International Politics*, Princeton, NJ, Princeton University Press.

Slater, A. E. (1996a), 'Can we afford to lose the pharmaceutical industry in the EU', *European Business Review*, 96, 4: 18–25.

Slater, A. E. (1996b), 'Recent developments in regulating the pharmaceutical business in the EU', *European Business Review*, 96, 1: 17–25.

Slater, A. E. (1998), 'The importance of the pharmaceutical industry to the UK economy', *Journal of Management in Medicine*, 12, 1: 5–20.

Smith, J. E. (1996), *Biotechnology* (3rd edition), Cambridge, Cambridge University Press.

Sorokin, D.Y. (1997), 'Combined microbial-chemical processes in the transformation of inorganic compounds: role in natural systems and possible applications in biotechnology', *Microbiology*, 66, 3: 241–8.

Spiers, E. M. (1994), *Chemical and Biological Weapons: A Study of Proliferation*, Basingstoke, Macmillan.

Steinberg, D. L. (1997), 'A most selective practice: the eugenic logics of IVF', in *Women's Studies International Forum*, 20, 1: 33–48.

Steinbrunner, J. D. (1998), 'Biological weapons: a plague upon all houses', *Foreign Policy*, Winter, 85–96.

Sterckx, S. (1997), *Biotechnology, Patents and Morality*, Aldershot, Ashgate.

Stilwell, M. (1999), 'Implications for developing countries of proposals to consider trade in genetically modified organisms (GMOs) at the WTO', CIEL Discussion Paper, http://www.twnside.org.sg/souths/twn/title/ciel-cn.html

Stock, T., M. Haug and P. Radler (1996), 'Chemical and biological weapon developments and arms control', *SIPRI Yearbook 1996*, Oxford, Oxford University Press/SIPRI, 661–708.

Stokke, O. (1997), 'Regimes as governance systems', in O. R. Young (ed.), *Global Governance: Drawing Insights From Environmental Experience*, Cambridge, MA, MIT Press, pp. 23–48.

Stopford, S. and S. Strange (with J. S. Henley) (1991), *Rival States, Rival Firms*, Cambridge, Cambridge University Press.

Strange, S. (1988), *States and Markets*, London, Pinter.

Strange, S. (1991), 'An eclectic approach', in C. N. Murphy and R. Tooze (eds), *The New International Political Economy*, Boulder, CO, Lynne Rienner, pp. 33–49.

Strange, S. (1994), *States and Markets* (2nd edition), London, Pinter.

Strange, S. (1996), *The Retreat of the State*, Cambridge, Cambridge University Press.

Süddeutsche Zeitung (1999), 'Mit Labor-Mais ist kein Geschäft zu machen', 4 September.

Sulej, J. C. (1998), 'UK international equity joint ventures in technology and innovation: an analysis of patterns of activity and distribution', *European Business Review*, 98, 1: 56–66.

Swanson, T. (1999), 'Why is there a biodiversity convention? The international interest in centralized development planning', *International Affairs*, 75, 2, April: 307–32.

Sylvan, R. (1985), 'A critique of deep ecology', *Radical Philosophy*, 40, Summer.

Takakura, Y. (1996), 'Development of drug delivery systems for macromolecular drugs', *Yakugaku Zasshi Journal of the Pharmaceutical Society of Japan*, 116, 7: 519–32.

Talalay, M., C. Farrands and R. Tooze (eds) (1997), *Technology, Culture and Competitiveness: Change and the World Political Economy*, London and New York, Routledge.

Taylor, C. C. W. (ed.) (1992), *Ethics and the Environment*, Oxford, Corpus Christi College.

Terry, J. and M. Calvert (eds) (1997), *Processed Lives: Gender and Technology in Everyday Life*, London, Routledge.

Thayer, A. M. (1999), 'Transforming agriculture', *Chemical and Engineering News*, 19 April, 21–35.

Thompson, M., R. Ellis and A. Wildavsky (1990), *Cultural Theory*, Oxford, Westview Press.

Thompson, P. B. (1997), *Food Biotechnology in Ethical Perspective*, London, Blackie.

Thumm, N. (1999), 'Patent protection for biotechnological inventions: incentive for European biotech innovators', *IPTS Report*, 33: 27–34.

Tinker, C. (1998), 'Responsibility for biological diversity conservation under international law', in C. Ku and P. F. Diehl (eds), *International Law: Classic and Contemporary Readings*, Boulder, CO, Lynne Rienner, pp. 415–442.

Tolba, M. K. *et al.* (1992), *The World Environment 1972–1992*, London, Chapman & Hall.

Tong, R. (1997), *Feminist Approaches to Bioethics: Theoretical Reflections and Practical Applications*, Boulder, CO, Westview Press.

Toro, T. (1992a), 'Revised gene law "panders to industry" while Germans flee paperwork', *New Scientist*, 135, 1833, 8 August, 6.

Toro, T. (1992b), 'Ease gene laws, say German MPs', *New Scientist*, 136, 1848, 21 November, 9.

Tóth, T. (1997), 'A window of opportunity for the BWC ad hoc group', *The CBW Conventions Bulletin*, Issue Number 37, September 1–5.

Tucker, J. (1998), 'Strengthening the BWC: Moving toward a compliance protocol', *Arms Control Today*, January/February.

Tucker, J. B. (1984–5), 'Gene wars', *Foreign Policy*, 57: 58–79.

Tucker, J. B. (1993), 'Lessons of Iraq's biological warfare programme', *Arms Control: Contemporary Security Policy*, 14, 3: 229–71.

Tucker, J. B. (1994), 'Dilemmas of dual-use technology: toxins in medicine and warfare', *Politics and the Life Sciences*, 13, 1: 51–62.

Tucker, J. B. (1997), 'The biological weapons threat', *Current History*, 97, 609: 167–72.

Tullis, E. C. *et al.* (1958), *The Importance of Rice and the Possible Impact of AntiRice Warfare*, Technical Study No. 5, Office of the Deputy Commander for Scientific Activities, Biological Warfare Laboratories, Fort Detrick, Maryland, March, 58–FDS–302.

Uddhav, K. and S. Ketan (1998), 'Advances in the human genome project: a review', *Molecular Biology Reports*, 25, 1: 27–43.

UN Centre on Transnational Corporations (1988), *Transnational Corporations in Biotechnology*, New York, United Nations.

Unattributable briefing (1996), Comment by an ex-US Government official at the conference on 'Allies Divided: US and European Perspectives on the Middle East', Harvard University, February.

UNCTAD (1997), *The World Investment Report 1997: Transnational Corporations, Market Structure and Competition Policy – Overview*, UNCTAD, Geneva.

Underdal, A. (1994), 'Measuring and explaining regime effectiveness', in H. Hveem (ed.), *Complex Cooperation: Institutions and Processes in International Resource Management*.

Union of Concerned Scientists (1999), 'Potential Harms to Health from Genetic Engineering', published at http://www.ucsusa.org/agriculture/gen.risks.health.html

United Nations (1969), 'Chemical and Bacteriological (Biological) Weapons and the Effects of Their Possible Use', A United Nations Report No. E.69.I.24 New York, Ballantine Books.

United Nations (1972), 'Convention on the Prohibition of the Development, Production and Stockpiling of Bacteriological (Biological) and Toxin Weapons and on their Destruction', United Nations General Assembly Resolution 2826 (XXVI), 16 December 1971.

United Nations (1986), 'Second Review Conference of the States Parties to the Convention on the Prohibition of the Development, Production and Stockpiling of Bacteriological (Biological) and Toxin Weapons and on their Destruction', Final Document Part II, Final Declaration, BWC/CONF.II/13/II, Geneva, 8–26 September.

United Nations (1991), 'Sixth Report Submitted by the Executive Chairman of the Special Commission (UNSCOM) to the Security Council pursuant to paragraph 9 (b) (I) of Security Council Resolution 687 (1991) of 3 April 1991', S/1998/920 (1998), 6 October 1998. Found at http://www.un.org/Depts/unscom/s98-920.html

United Nations (1992), 'Third Review Conference of the States Parties to the Convention on the Prohibition of the Development, Production and Stockpiling of Bacteriological (Biological) and Toxin Weapons and on their Destruction', Final Document, BWC/CONF.III/23, Geneva, 9–27 September.

United Nations (1994), 'Special Conference of the States Parties to the Convention on the Prohibition of the Development, Production and Stockpiling of Bacteriological (Biological) and Toxin Weapons and on their Destruction' BWC/SPCONF/I, Geneva 19–30 September, http://www.brad.ac.uk/acad/sbtwc/verex/verex1.htm

United Nations (1995a), 'Convention on Biological Diversity', Second Conference.

United Nations (1995b), United Nations Environment Programme (UNEP), 'UNEP International Technical Guidelines for Safety in Biotechnology', UNEP, Nairobi, Kenya, 1 (Introduction).

United Nations (1996), 'Fourth Review Conference of the States Parties to the Convention on the Prohibition of the Development, Production and Stockpiling of Bacteriological (Biological) and Toxin Weapons and on their Destruction', Final Document, BWC/CONF.IV/9, Geneva, 25 November–6 December.

United Nations (1997), 'Plant pathogens important for the BWC', Working Paper by South Africa, BWC/Ad Hoc Group/WP.124, 3 March.

United Nations (1998), 'Working Paper submitted by the United Kingdom of Great Britain and Northern Ireland on behalf of the European Union', Ad Hoc Group of States Parties to the Convention on the Prohibition of the Development, Production and Stockpiling of Bacteriological (Biological) and Toxin Weapons and on their Destruction, BWC/AD HOC GROUP.WP.272, 9 March.

United Nations (1999), 'Working Paper submitted by the Netherlands and New Zealand', Ad Hoc Group of States Parties to the Convention on the Prohibition of the Development, Production and Stockpiling of Bacteriological (Biological) and Toxin Weapons and on their Destruction, BWC/AD HOC GROUP/WP.362, 6 April.

United Nations Development Programme (1998), *Human Development Report 1998*, Oxford, Oxford University Press.

US Congress (1970), 'Chemical-Biological Warfare: U.S. Policies and International Effects', *Hearings before the Subcommittee on National Security and Scientific Developments of the Committee on Foreign Affairs House of Representatives, ninety-first Congress, First Session, 18, 20 November; 2, 9, 18, and 19 December 1969* Washington, DC, US Government Printing Office.

US Congress (1993a), Office of Technology Assessment, 'Proliferation of Weapons of Mass Destruction: Assessing the Risks', OTA-ISC-559, Washington, DC, US Government Printing Office, August.

US Congress (1993b), Office of Technology Assessment, 'Technologies Underlying Weapons of Mass Destruction', OTA-BP-ISC-115, Washington, DC, US Government Printing Office, December.

US Congress, Office of Technology Assessment (1991), 'Biotechnology in a global economy', OTA-BA-494, Washington, DC, US Government Printing Office.

van der Plank, J. E. (1963), *Plant Diseases: Epidemics and Control*, New York and London, Academic Press.

Van Tulder, R. and G. Junne (1988), *European Multinationals in Core Technologies*, Chichester, New York, John Wiley & Sons.

Vavilov, N. I. (1951), 'The origin, variation, immunity and breeding of cultivated plants', *Chronica Botanica*, 1, 3: 1–364.

Vogel, D. (1995), *Trading Up. Consumer and Environmental Regulation in a Global Economy*, Cambridge, MA, Harvard University Press.

Vogler, J. (1981), 'Technology and change in international relations: on the independence of a variable', in B. Buzan and R. J. B. Jones (eds), *Change and the Study of International Relations: The Evaded Dimension*, London, Pinter.

Vogler, J. (1995), *The Global Commons: A Regime Analysis*, Chichester, John Wiley.

von Weizsacker, C. (1993), 'Competing notions of biodiversity', in W. Sachs (ed.), *Global Ecology*, London, Zed Books, pp. 117–31.

von Weizsacker, E. U. (1994), *Earth Politics*, London, Zed Books.

Walsh, V. and I. Galimberti (1993), 'Firm strategies, globalisation and new technological paradigms: The case of biotechnology', in M. Humbert (ed.) *The Impact of Globalisation on Europe's Firms and Industries*, London, Pinter, pp. 175–90.

Walter, C., S. D. Carson, M. Menzies, T. Richardson, and M. Carson (1998), 'Application of biotechnology to forestry: molecular biology of conifers', *World Journal of Microbiology and Biotechnology*, 14, 3: 321–30.

Waltz, K. N. (1979), *Theory of International Politics*, Reading, MA, Addison-Wesley Publishing Company.

Wapner, P. (1997), 'Governance in global civil society', in O. R. Young (ed.), *Global Governance: Drawing Insights from the Environmental Experience*, Cambridge, MA, MIT Press, pp. 65–84.

Warner, D. (1992), *An Ethic of Responsibility in International Relations*, London, Lynne Rienner.

Washington Post (1999), 'Biotech crops spur warning: 30 farm groups say consumer backlash could cost markets', 24 November.

Wateringen, S. van de (1997), 'US pushes Ecuador to sign IPR agreement', *Biotechnology and Development Monitor*, No.33, December, 20–2.

Watson, J. D. and G. S. Stent (1998*), The Double Helix: A Personal Account of the Discovery of the Structure of DNA*, Simon & Schuster.

WCED 1987, *Our Common Future* (The Brundtland Report), Oxford, Oxford University Press.

Wechsler, J. (1998), 'Washington report – pressures and priorities', *Pharmaceutical Technology Europe*, 10, 10: 18–27.

Welles, E. O. (1995), 'The Awakening', *Inc Online*, 23, sourced, April 1999, from http://www.inc.com/incmagazine/archives/01950231.html

Wells, A. J. (1994), 'Patenting new life forms: an ecological perspective', *EIPR*, 6, 3: 111–18.

Wells, D. and J. K. Sherlock (1998), 'Strategies for preimplantation genetic diagnosis of single gene disorders by DNA amplification', *Prenatal Diagnosis*, 18, 13: 1389–401.

Werner, R. G. (1998), 'The value of contract manufacturing', *Pharmaceutical Technology Europe*, 10, 10: 60–71.

Westing, A. H. (1985), 'The threat of biological warfare', *Bioscience*, 35, 10, November.

Whatmore, S. and L. Thorne (1997), 'Nourishing networks: alternative geographies of food', in D. Goodman and M. J. Watts (eds), *Globalising Food*, London, Routledge, pp. 287–304.

Wheale, P. R. and R. M. McNally (1993), 'Biotechnology policy in Europe: a critical evaluation', *Science and Public Policy*, 20, 4, August: 261–79.

Wheale, P. R. and R. M. McNally (1994), 'What 'bugs' genetic engineers about bioethics: The consequences of genetic engineering as a post-modern technology', in Dyson and Harris (1994).

Wheelis, M. (1997), 'Addressing the Full Range of Biological Warfare in a BWC Compliance Protocol', Paper presented at Pugwash Meeting No. 229, 20–1 September, Geneva, Switzerland.

Wheelis, M. (1999), 'Outbreaks of disease: current disease reporting', Briefing Paper No. 21, April.

Whitby, S. and P. Rogers (1997), 'Anti-crop biological warfare – implications of the Iraqi and US Programs', Defense Analysis, 13, 3: 303–18.

Whitley, R. (1994), 'The comparative analysis of business systems', in R. Whitely (ed.), *European Business Systems*, London, Sage, pp. 5–46.

Wieandt, A. and N. Amin (1994), 'Biotechnology: the emerging battlefield for US and Japanese pharmaceutical companies', *Technology Analysis & Strategic Management*, 6, 4: 423–35.

Wiegele, T. C. (ed.) (1990), 'Biotechnology and international conflict', Special edition of *Politics and the Life Sciences*, 9, 1.

Wiegele, T. C. (1991*), Biotechnology and International Relations: The Political Dimensions*, Gainesville, University of Florida Press.

Wijk, J. van (1992), 'GATT and the legal protection of plants in the Third World', *Biotechnology and Development Monitor*, no.10, March.

Williams, O. (1998a), 'Sui generis rights: a balance misplaced', *Signposts to Sui*

Generis, Barcelona, BioThai and Grain.

Williams, O. (1998b), 'Trade and intellectual property regimes', Conference paper presented at the 1998 International Political Economy (IPEG) Workshop, 25–27 October, Liverpool John Moores University.

Wilmut, I., A. E. Schnieke, J. McWhir, A. J. Kind and K. H. S. Campbell (1997a), 'Viable offspring derived from fetal and adult mammalian cells', *Nature*, 385, 6619: 810–13.

Wilmut, J., J. McWhir, and K. H. S. Campbell (1997b), 'Cloned sheep: implications for medicine and biological research', *Human Reproduction*, 12, 5: S10.

Winner, L. (1977), *Autonomous Technology: Technics Out of Control as a Theme in Political Thought*, Cambridge, MA, MIT Press.

Wolf, O. (1999), 'Transatlantic investments and human capital formation: the case of biotech firms', *IPTS Report*, 33: 34–9.

Wood and Fairley (1998), *Chemical Week*, 4–11 February, 27.

Working Paper Submitted by South Africa (1997) 'Plant pathogens of importance to the BWC', BWC/Ad Hoc Group/WP.124, March.

World Health Organization (1970), *Health Aspects of Chemical and Biological Weapons*, Geneva, WHO, November.

Wright, S. (1994), *Molecular Politics*, Chicago, University of Chicago Press.

Wu, A. H., M. C. Pike and D. O. Stram (1999), 'Meta-analysis: dietary fat intake, serum estrogen levels, and the risk of breast cancer', *Journal of the National Cancer Institute*, 91, 6: 529–34.

Wyatt-Walter, A. (1995), 'Globalization, corporate identity and technology policy', *Journal of European Public Policy*, 2, 3: 437–45.

Wynne, B. (1999), 'Bitter fruits', *Guardian*, G2, 16 September.

Young, O., P. Stern and D. Druckman (eds) (1992), *Global Environmental Change: Understanding the Human Dimensions*, Washington, DC, National Academy Press.

Youngs, G. (1997), 'Culture and the technological imperative: missing dimensions', in M. Talalay, C. Farrands and R. Tooze (eds), *Technology, Culture and Competitiveness*, London and New York, Routledge, pp. 27–40.

Yuan, R. T. (1987), *Biotechnology in Western Europe*, Washington, DC, US Department of Commerce, International Trade Administration.

Yuval-Davis, N. (1997), *Gender and Nation*, London, Sage.

Zilinskas, R. A. (1995) 'UNSCOM and the UNSCOM experience in Iraq', in 'Symposium of United Nations biological weapons inspectors: implications of the Iraqi experience for biological arms control', special issue of *Politics and the Life Sciences*, 14, 2: 230–5.

Zilinskas, R. A. (1998), 'Verifying compliance to the Biological and Toxin Weapons Convention' in R. M. Atlas (ed.), *Critical Reviews in Microbiology: Special Issue Biological Weapons*, 34, 3: 195–218.

Index

Note: 'n' after a page reference refers to a note number on that page.